S0-BZI-765

Ecotoxicology in Theory and Practice

Chapman & Hall Ecotoxicology Series

Series Editors

Michael H. Depledge
Professor of Ecotoxicology, Institute of Biology, Odense University, Denmark

Brenda Sanders
Associate Professor of Physiology, Molecular Ecology Institute, California State University, USA

In the last few years emphasis in the environmental sciences has shifted from direct toxic threats to humans, towards more general concerns regarding pollutant impacts on animals and plants, ecosystems and indeed on the whole biosphere. Such studies have led to the development of the scientific discipline of ecotoxicology. Throughout the world socio-political changes have resulted in increased expenditure on environmental matters. Consequently, ecotoxicological science has developed extremely rapidly, yielding new concepts and innovative techniques that have resulted in the identification of an enormous spectrum of potentially toxic agents. No single book or scientific journal has been able to keep pace with these developments.

This series of books provides detailed reviews of selected topics in Ecotoxicology. Each book includes both factual information and discussions of the relevance and significance of the topic in the broader context of ecotoxicological science.

Already published

Animal Biomarkers as Pollution Indicators
David B. Peakall
Hardback (0 412 40200 9), 292 pages

Ecotoxicology in Theory and Practice

V.E. Forbes and T.L. Forbes
Department of Marine Ecology and Microbiology,
National Environmental Research Institute,
Denmark

CHAPMAN & HALL
London · Glasgow · New York · Tokyo · Melbourne · Madras

Published by Chapman & Hall, 2–6 Boundary Row, London SE1 8HN

Chapman & Hall, 2–6 Boundary Row, London SE1 8HN, UK

Blackie Academic & Professional, Wester Cleddens Road, Bishopbriggs, Glasgow G64 2NZ, UK

Chapman & Hall Inc., One Penn Plaza, 41st Floor, New York NY 10119, USA

Chapman & Hall Japan, Thomson Publishing Japan, Hirakawacho Nemoto Building, 6F, 1–7–11 Hirakawa-cho, Chiyoda-ku, Tokyo 102, Japan

Chapman & Hall Australia, Thomas Nelson Australia, 102 Dodds Street, South Melbourne, Victoria 3205, Australia

Chapman & Hall India, R. Seshadri, 32 Second Main Road, CIT East, Madras 600 035, India

First edition 1994

Typeset in 10/12pt Times by ROM-Data Corporation Ltd, Falmouth
Printed in Great Britain by T J Press (Padstow) Ltd, Padstow, Cornwall

ISBN 0 412 43530 6

A catalogue record for this book is available from the British Library

Library of Congress Cataloging-in-Publication data available

∞ Printed on permanent acid-free text paper, manufactured in accordance with the proposed ANSI/NISO Z 39.48–199X and ANSI Z 39.48–1984

Contents

Series foreword

Ecotoxicology is a relatively new scientific discipline. Indeed, it might be argued that it is only during the last 5–10 years that it has come to merit being regarded as a true science, rather than a collection of procedures for protecting the environment through management and monitoring of pollutant discharges into the environment. The term 'ecotoxicology' was first coined in the late sixties by Prof. Truhaut, a toxicologist who had the vision to recognize the importance of investigating the fate and effects of chemicals in ecosystems. At that time, ecotoxicology was considered a sub-discipline of medical toxicology. Subsequently, several attempts have been made to portray ecotoxicology in a more realistic light. Notably, both Moriarty (1988) and F. Ramade (1987) emphasized in their books the broad basis of ecotoxicology, encompassing chemical and radiation effects on all components of ecosystems. In doing so, they and others have shifted concern from direct chemical toxicity to humans, to the far more subtle effects that pollutant chemicals exert on natural biota. Such effects potentially threaten the existence of all life on earth.

Although I have identified the sixties as the era when ecotoxicology was first conceived as a coherent subject area, it is important to acknowledge that studies that would now be regarded as ecotoxicological are much older. Wherever people's ingenuity has led them to change the face of nature significantly, it has not escaped them that a number of biological consequences, often unfavourable, ensue. Early waste disposal and mining practices must have alerted the practitioners to effects that accumulated wastes have on local natural communities; for example, by rendering water supplies undrinkable or contaminating agricultural land with toxic mine tailings. As activitites intensified with the progressive development of human civilizations, effects became even more marked, leading one early environmentalist, G.P. Marsh, to write in 1864: 'The ravages committed by Man subvert the relations and destroy the balance that nature had established'.

But what are the influences that have shaped the ecotoxicological studies of today? Stimulated by the explosion in popular environmentalism in the sixties, there followed in the seventies and eighties a tremendous increase in the creation of legislation directed at protecting the environment. Furthermore, political restructuring, especially in Europe, has led to the widespread implementation of this legislation. This currently involves enormous numbers of environmental managers, protection officers, technical staff and consultants. The ever-increasing use of new chemicals places further demands on government agencies and industries who are required by law to evaluate potential toxicity and likely environmental impacts. The environmental manager's problem is that he needs rapid answers to current questions concerning a very broad range of chemical effects and also information about how to control discharges, so that legislative targets for *in situ* chemical levels can be met. It is not surprising, therefore, that he may well feel frustrated by more research-based ecotoxicological scientists who constantly question the relevance and validity of current test procedures and the data they yield. On the other hand, research-based ecotoxicologists are often at a loss to understand why huge amounts of money and time are expended on conventional toxicity testing and monitoring programmes, which may satisfy legislative requirements, but apparently do little to protect ecosystems from long-term, insidious decline.

It is probably true to say that until recently ecotoxicology has been driven by the managerial and legislative requirements mentioned above. However, growing dissatisfaction with laboratory-based tests for the prediction of ecosystem effects has enlisted support for studying more fundamental aspects of ecotoxicology and the development of conceptual and theoretical frameworks.

Clearly, the best way ahead for ecotoxicological scientists is to make use of the strengths of our field. Few sciences have at their disposal such a well-integrated input of effort for people trained in ecology, biology, toxicology, chemistry, engineering, statistics, etc. Nor have many subjects such overwhelming support from the general public regarding our major goal: environmental protection. Equally important, the practical requirements of ecotoxicological managers are not inconsistent with the aims of more academically-orientated ecotoxicologists. For example, how better to validate and improve current test procedures than by conducting parallel basic research programmes *in situ* to see if controls on chemical discharges really do protect biotic communities?

More broadly, where are the major ecotoxicological challenges likely to occur in the future? The World Commission on Environment and Development estimates that the world population will increase from *c.* 5 billion at present to 8.2 billion by 2025. 90% of this growth will occur in developing countries in subtropical and tropical Africa, Latin America and Asia. The introduction of chemical wastes into the environment in these regions is likely to escalate dramatically, if not due to increased industrial output, then due to the use of pesticides and fertilizers in agriculture and the disposal of damaged, unwanted or obsolete consumer goods supplied from industrialized countries. It may be many years before resources become available to implement effective waste-recycling programmes in countries with poorly developed infrastructures, constantly threatened by natural disasters and poverty. Furthermore, the fate, pathways and effects of chemicals in subtropical and tropical environments have barely begun to be addressed. Whether knowledge gained in temperate ecotoxicological studies is directly applicable in such regions remains to be seen.

The Chapman & Hall Ecotoxicology Series brings together expert opinion on the widest range of subjects within the field of ecotoxicology. The authors of the books have not only presented clear, authoritative accounts of their subject areas, but have also provided the reader with some insight into the relevance of their work in a broader perspective. The books are not intended to be comprehensive reviews, but rather accounts which contain the essential aspects of each topic for readers wanting a reliable introduction to a subject or an update in a specific field. Both conceptual and practical aspects are considered. The Series will be constantly added to and books revised to provide a truly contemporary view of ecotoxicology. I hope that the Series will prove valuable to students, academics, environmental managers, consultants, technicians, and others involved in ecotoxicological science throughout the world.

Michael Depledge
Odense, Denmark

Preface

Why should we tolerate a diet of weak poisons, a home in insipid surroundings, a circle of acquaintances who are not quite our enemies, the noise of motors with just enough relief to prevent insanity? Who would want to live in a world which is just not quite fatal?

Paul Shepard

We decided to write this book because we felt that there was a desire among ecotoxicologists to assess the current approaches used in the study of pollutants and to evaluate the tortuous path followed by new scientific knowledge as it is translated into the legislation that is presumed to protect our environment. As both of us are relatively new to the field of ecotoxicology, we felt we were in an ideal position to provide a fresh and relatively unbiased analysis of some of the currently accepted (and some not so accepted) principles and practices of ecotoxicology. Many readers may disagree. But as one of our primary aims is to encourage debate on some of the more controversial aspects associated with the science and application of ecotoxicology, we hope that, to some extent, you will disagree – with us and with each other – because that is a sign of a healthy scientific enterprise.

Constructive debate on topics of general concern requires an extensive degree of communication among parties. Unfortunately this has not always been the case in ecotoxicology. Thus in an attempt to facilitate communication among those involved in the different aspects of the science and its application, we have sought the expertise and opinions of many people involved in different areas of ecotoxicology. Our goal was to present a balanced, though not exhaustive, view of some important topics in order to promote cooperation and understanding among those working in academia, regulatory agencies, private testing firms and industry.

With the extremely rapid and apparently 'unbrakeable' evolution of contemporary economic and industrial development, ecotoxicologists are under great pressure to develop new and improved methods to assess and predict man's impact on the environment. Such pressure can make it difficult for scientists, legislators and managers to spare the time and energy to step back and evaluate the general implications of their work and where it is headed. But unless this is done, our efforts will be of little long-term value to either science or society at large.

Given the broad range in experience and areas of expertise of those involved in contemporary ecotoxicology, we have generally assumed no more than basic university-level training in the natural sciences in an attempt to address both students and practicing professionals in ecotoxicology. Nevertheless, one caveat is in order. The problems faced by ecotoxicologists are difficult. Some of the current research areas in the field pose questions to scientists that will be as tough to answer as any the natural world has to offer. In many areas of ecotoxicology we may not yet know enough even to ask the right questions. Because of the inherent difficulty of their field, ecotoxi cologists need to feel comfortable dealing with research questions at the appropriate level of complexity. Thus we have included in the form of Chapter 5 some material that is, even when pared to the bare minimum, more detailed and complex than that of the other chapters. Adapting the words of evolutionary biologist Douglas Futuyma to the present context – there is no reason to expect ecotoxicology books to be easier than those in physics and chemistry, disciplines whose subject matter is so much simpler.

The first chapter focuses on the definition and scope of modern ecotoxicology. Because the definition of ecotoxicology will tend to constrain the areas of research, we felt it important to discuss the definition of the science from an historical perspective. This allows us to develop an appreciation of the evolution of current research questions and methods. In the second chapter we look at the role of nonscientific issues in ecotoxicology. Given ecotoxicology's inherently applied nature, nonscientific considerations often play a major role in the shaping of research questions and distribution of funds. Ecotoxicologists need to appreciate and understand this role. In this chapter we also take a closer look at environmental management and risk assessment. Chapter 3 examines the methods currently used to study pollutant effects on populations, communities and ecosystems in both the field and laboratory. The fourth chapter investigates the fundamental dose – response relationship in detail. We examine some of the limits of its application and discuss the importance of obtaining results that are as precise and accurate as currently available statistical methods allow. We also discuss the nature and proper interpretation of the variability that is always observed when conducting toxicity texts.

There has been much recent discussion within ecotoxicological circles concerning the need to deal with higher levels of biological organization and for integrated studies linking chemical fate and effect in the environment. Chapter 5 represents our attempt to come to grips with this very challenging problem. In order to demonstrate the linkages and interactions clearly, we have taken a very close look at a specific but very important area – the fate and effect of organic pollutants in aquatic sediments. In doing so we have become convinced that further progress in other areas will require at least this level of detail. Finally, in chapter 6 we draw together the achievements, knowledge gaps and future challenges within the field of ecotoxicology and offer suggestions concerning the approaches to be taken in the future.

Acknowledgements

This book has been improved immeasurably by the generous assistance of many people. Deserving special mention are the patient reviewers of the initial drafts of each of the chapters. We thank Finn Bro-Rasmussen, Tom Fenchel, Merete Reuss (Chapter 1); John Gray, Henrik Tyle (Chapter 2); Hans Løkke, Charles H. Peterson (Chapter 3); Donald Baird, Leon D. Forbes, John Gray, Søren Pedersen (Chapter 4); Kim Gustavson, Frank Stuer-Lauridsen (Chapter 5); Peter Calow, Jeffrey Levinton (Chapter 6) and Liv Kure (Chapters 1, 2, 3 and 5) for generously supplying their time and energy. We would also like to thank Finn Bro-Rasmussen, Axel Damborg, Michael Depledge, Preben Kristensen and Henrik Tyle for critical discussions of the nature and scope of ecotoxicological research which helped to crystallize some of the opinions expressed in this book. Henrik Tyle provided ready access to a seemingly endless stream of grey literature. Jean Massey supplied written information on the toxicological testing requirements faced by industry. Tove Juul Hansen and the Risø library staff provided superb access to a very large and diverse literature. Michael Depledge and Leon D. Forbes read and commented on the entire manuscript.

We thank the Biology Institute at Odense University for the use of facilities during manuscript preparation. Denmark's National Environmental Research Institute provided a most collegial working environment and the freedom to pursue the writing. Partial support from the Carlsberg Foundation (VF) and the Danish National Science Council (TF) is gratefully acknowledged.

1 *Answers to some important questions*

We need to do something in haste
About the production of waste
For if we do not
Then what have we got
But a world that is not to our taste.
Kenneth E. Boulding, 1990

Human activities have dramatically altered this planet and its inhabitants. While recognizing that many of our activities have had undesirable effects, we appear willing to accept some degree of harm as the price of progress. But how can the degree of harm be measured and how much should be tolerated? While the appalling levels of pollution in our big cities are often the subject of public outrage, the production, consumption and disposal of ever-increasing quantities of new chemicals continues. We have little idea of how the myriad chemicals present in our air, water, soil and food affect us and the other species with which we share the Earth. Although we have undeniably achieved some degree of 'better living through chemistry', we now realize that continued development involves formidable tradeoffs. We have only a limited idea of what these tradeoffs might be and how to quantify them. We therefore hire experts to study the effects of our activities, to estimate the risk of such activities to our health and environment and, should some of these risks prove unacceptable, to protect us from the effects of the chemicals we have made and the wastes we have produced. These experts are called 'ecotoxicologists'. This book is about what they do and how they may be able to do it better.

In the following chapters, we discuss some current approaches to the study of chemical pollutants, emphasizing in particular those areas that we have found to be the subject of greatest confusion or controversy. Our goal in presenting these issues is to clarify important concepts and questions in ecotoxicology in order to highlight promising directions for future research.

We begin by attempting to define the scope and objectives of ecotoxicology. We look at how the field was originally conceived, how it is viewed today and how it is pursued in practice.

1.1 What is ecotoxicology?

1.1.1 A hodgepodge of definitions

As its name implies, ecotoxicology merges the fields of ecology and toxicology. However, this merger is characterized by a considerable amount of confusion and inconsistency. Toxicology has been defined as 'the study of the adverse effects of chemicals on living organisms' (Klaassen and Eaton, 1991). Toxicology is generally concerned with the effects of chemicals at the level of the individual organism, or its constituent parts, and emphasizes the mechanistic bases of harmful effects and the conditions under which they occur (American Institute of Biological Sciences, 1978). Some authors restrict the scope of toxicology to the study of xenobiotic chemicals (Gallo and Doull, 1991). But what is a xenobiotic chemical? Sipes and Gandolfi (1991) defined a xenobiotic as a chemical that is foreign to the human or another organism's body. So defined, xenobiotics can be either natural or artificial. In contrast, Rand and Petrocelli (1985) defined a xenobiotic as 'A foreign chemical or material not produced in nature and not normally considered a constitutive component of a specified biological system. This term is usually applied to manufactured chemicals.' Then there are those, such as Klaassen and Rozman (1991) who have explicitly used the terms toxicant, drug, xenobiotic and chemical interchangeably.

The original definition of ecotoxicology is generally attributed to Truhaut (1977) who defined ecotoxicology as 'the branch of Toxicology concerned with the study of toxic effects, caused by natural or synthetic pollutants, to the constituents of ecosystems, animal (including human), vegetable and microbial, in an integral context.' Since 1977, ecotoxicology has acquired a number of definitions, reflecting the fact that this area of study is still in its infancy and that its students are not entirely agreed on its direction and scope. A perusal of the literature of the last decade demonstrates the differences of opinion as to what ecotoxicologists study. Moriarty (1983) defined ecotoxicology as the study of the effects of pollutants on ecosystems. Butler (1984) considered ecotoxicology 'a study of the effects of released pollutants on the environment and on the biota that inhabit it.' Hayes (1991) has recently added to the confusion by defining environmental toxicology as 'all aspects of toxicology that emphasize organisms other than humans and their domestic animals' and ecological toxicology as 'the study of all toxicants produced by living organisms and of the ecological relationships made possible by these poisons.' In contrast, Klaassen and Eaton (1991) have contended that environmental toxicology 'focuses on the impacts of chemical pollutants found in the environment on biological organisms', and they defined ecotoxicology as 'a specialized area within environmental toxicology that focuses more specifically on the impacts of toxic substances on population dynamics within an ecosystem.' The question as to what would constitute a nonbiological organism aside, the above definitions demonstrate the confusion that exists over what constitutes the discipline of ecotoxicology.

Some workers have focused on specific groups of organisms and have defined fields such as aquatic toxicology – 'the qualitative and quantitative study of the adverse or toxic effects of chemicals and other anthropogenic materials or xenobiotics on aquatic organisms ... Aquatic toxicology is also concerned with the concentrations or quantities of chemicals that can be expected to occur in the aquatic environment in water, sediment, or food. Therefore, it includes the study of the transport, distribution, transformation and ultimate fate of chemicals in the aquatic environment' (Rand and Petrocelli, 1985). Aquatic toxicology, as defined by Rand and Petrocelli, is not restricted to toxicological studies of aquatic organisms, but deals in fact with ecotoxicology of aquatic ecosystems. Thus the term aquatic ecotoxicology would more accurately describe studies whose goals meet the above definition.

It should be noted in this context that Neuhold (1986) claimed that aquatic toxicology emerged in the mid-1920s as a science for testing the effects of organic chemicals on aquatic organisms. If restricted to the **toxicology** of aquatic species, this field can be considered to precede the birth of ecotoxicology. Interested readers should consult Gallo and Doull (1991) for a concise summary of the history and development of toxicology.

Although Moriarty, an ecologist, (1983) defined ecotoxicology as a subdiscipline of ecology, it is clear that toxicologists, such as Truhaut (1977), Hayes (1991), Klaasen and Eaton (1991), have considered ecotoxicology as a subdiscipline of toxicology. More recently, ecotoxicology has come to be regarded as a new field in its own right, presumably consisting of a relatively self-contained research programme not subsumed by related disciplines (Butler, 1984).

Partly because the scope of ecotoxicology has not been consistently defined, the field often seems to suffer from a confusion of goals and a lack of focus. For example, the 1992 European Conference on Ecotoxicology (SECOTOX) included a session entitled 'ecological ecotoxicology'. The problem of redundancy aside, this title implies that the bulk of ecotoxicology is ecologically lacking. We have argued elsewhere that this is indeed true (Chapters 4 and 6). In any case, at this meeting only nine out of 128 talks were included in the ecological ecotoxicology session. Presumably the remaining 119 talks, or 92% of the meeting presentations, fit into the absurd category of non-ecological ecotoxicology.

By taking a closer look at the historical influences that have shaped the development of ecotoxicology we may be able to increase our understanding of where the field is today and where it may be headed in the future. We begin by setting the stage with the cultural–political scene that facilitated the birth of ecotoxicology, and proceed by briefly tracing the origin and historical development of the field.

1.1.2 Historical perspective

The level of governmental attention focused on environmental problems began to increase sharply during the 1950s (Squires, 1983). In 1954, an international meeting was held in London which formulated the International Convention for the Prevention of Pollution of the Sea by Oil. The Convention of the High Seas, held in 1958, the Convention on the Territorial Sea and the Contiguous Zone and the Convention on the Continental Shelf (both adopted by the USA in 1964) represented important achievements of the international community to reduce pollution. Legislative activities within European and North American countries devoted increasing attention to the need for pollution control (Squires, 1983). For example, by 1961, USA law prohibited the discharge of oil in American coastal waters. In 1972, the American Ocean Dumping Act was passed, and in 1973 several amendments to the Water Quality Act were passed. In 1976, President Ford signed the Toxic Substances Control Act. This served to expand the role that ecotoxicological studies would play in American science. With the signing of this law, the USA Environmental Protection Agency could begin to require premarket notification from the manufacturer regarding the safety of new chemicals (Cairns *et al.*, 1978). This in turn provided an effective stimulus for the development of testing protocols and hence to the advent of the field of environmental risk assessment. On the international front, the Commission of the European Communities and the Organization for Economic Cooperation and Development (OECD) prompted, and continue to provide, an important framework and stimulus for the standardization of toxicity testing and risk assessment procedures.

The coining of **ecotoxicology** by toxicologists in the late 1960s signalled an intensification of public awareness that man's increasing reliance on chemicals could have undesirable consequences for his environment and his health. This awareness was provoked by an accumulating number of chemical-related incidents which were vividly captured in such books as Rachel Carson's *Silent Spring* (1962).

According to Truhaut (1977), the word ecotoxicology was coined during a meeting held in June of 1969, by an ad-hoc committee of the International Council of Scientific Unions, of which he was a member. Truhaut's definition was initially embraced by toxicologists who recognized the importance of interactions among artificial chemicals, the effects of those chemicals on the environment (especially on the living components) and the consequences of the environmental release of chemicals for human health. Although Truhaut (1977) acknowledged the importance of ecology as a component of ecotoxicology, it is clear that his main concern was toxicological (and his ultimate concern was with human heath). His concept of an ecosystem as a 'supraspecific' organism (though he did not define the term) is not one likely to receive much support from modern ecologists (Tilman, 1989). [As we discuss in Chapter 3, many ecologists continue to use individual and ecosystem 'health' as analogies. This practice does little to clarify the distinction between an organism and an ecosystem.]

The concept of an **ecosystem** was first proposed by Tansley in 1935 (Schlesinger, 1989). The branch of ecology that deals with the study of 'the circulation, transformation, and accumulation of energy and matter through the medium of living things and their activities' is generally known as **systems** or **ecosystems** ecology (Evans, 1956, cited in Schlesinger, 1989). Although Truhaut (1977) identified the study of the fate and cycling of pollutants in ecosystems as an important principle in ecotoxicological studies, many ecotoxicologists appear to be unaware of the extent to which the principles of systems ecology can contribute to the study of ecotoxicological problems. This is unfortunate, as among the many insights that systems ecology can provide into the study of ecotoxicological problems, is the recognition that the fate and effects of substances in ecosystems are inseparable components of ecosystem function (Chapter 5 demonstrates this point).

Even though toxicologists can be credited with defining the new field of ecotoxicology in the late 1960s, ecologists had, in fact, already been studying the effects of chemical pollutants on living systems for about 40 years. At least as early as the 1930s ecologists were investigating the effects of industrialization on natural populations of plants and animals. For example, Prát published investigations of metal tolerance in plants occupying metal contaminated habitats in 1934 (cited in Moriarty, 1983). During the 1960s, studies of the grass *Agrostis tenuis* colonizing soils around heavy metal mines examined the interplay between selection and gene flow in the evolution of adjacent plant populations (McNeilly and Bradshaw, 1968; McNeilly, 1968). Studies performed in 1932 in Denmark, and in 1959 in the USA, documented decreases in species abundance and diversity in polluted habitats (Klerks and Levinton, 1989). These early studies of artificial changes provided valuable opportunities for ecologists because the relatively great speed and magnitude of such changes facilitated the study of ecological and evolutionary processes. For example, studies on the control of agricultural pests have contributed substantially to our understanding of the basic factors influencing population dynamics of natural populations (Ricklefs, 1979). Although these early studies were perhaps not inspired by the same environmental fervour that spurred the birth of ecotoxicology, such work, in retrospect, can nevertheless contribute a great deal to our current understanding of ecotoxicological phenomena. Also, recent work on the ecological and evolutionary responses of biota to stress is directly relevant to understanding the biological consequences of pollution. Two excellent books on this subject are *The Effects of Stress and Pollution on Marine Animals* edited by Bayne *et al.*, 1985 and *Evolution, Ecology and Environmental*

Stress edited by Calow and Berry, 1989. Many of the approaches used to test so-called ecotoxicity have been heavily criticized for their lack of ecological relevance, and we count ourselves among those who feel that ecotoxicology is in serious need of a more ecological perspective.

Ecotoxicologists must be conversant in the languages of both ecology and toxicology. The inherent nature of ecotoxicological problems is such that research questions must also consider physical and chemical principles in relation to phenomena occurring at different levels of biological organization. Much more than most other scientific disciplines, research problems in ecotoxicology are shaped by social, political and economic considerations in addition to scientific principles. How can all of these components be fitted together so that the scope and direction of ecotoxicology can be unambiguously defined?

1.1.3 Linking ecology and toxicology

In a 1986 book review of *Multispecies Toxicity Testing* edited by J. Cairns, Jr., Brungs lamented that 'The fields of toxicology and ecology are rarely associated, either academically or professionally. (There are efforts to create an ecotoxicological science, but its impact has yet to be demonstrated.).' With regard to ecologists and toxicologists, Cairns (1990) stated that 'it would be hard to find two groups with less interchange than these.' In a critique of modern approaches to the study of ecotoxicology, Moriarty (1983) wrote, 'On occasion, the only difference between toxicology and ecotoxicology appears to be in the species selected for toxicological tests: acute toxicity is measured on the water flea instead of the laboratory rat.' In this context, it is fascinating to note that in the base data set, required before substances are manufactured or marketed in the European Community (sixth amendment of directive 67/548/EEC), tests of acute toxicity to mammals are defined as **toxicological studies** and tests of acute toxicity to fish and crustaceans as **ecotoxicological studies** (Commission of the European Communities, 1989). Although it is possible to explain the toxicological bias in ecotoxicology as an historical consequence of its development, future advances in the field clearly require a more even balance between ecology and toxicology.

Our ultimate interest as ecotoxicologists must centre on determining the effects of pollutants on the structure and function of intact ecosystems, communities or assemblages. However, the complexity at this level of biological organization has generally precluded direct measurements of effects on natural ecosystems and has directed study toward separate components making up the system and transfers of matter and energy among them. This has been done in the hope of assembling the whole from its parts. Clearly, achieving a thorough understanding of the effects of chemicals on living systems benefits from studies at many different levels of biological organization. For example, it is not sufficient to know that exposure to organotin reduces the recruitment of some gastropod population into an estuary. It is also beneficial to know the mechanism behind such reduction and the effect of such reduction on population dynamics for this species. Is larval or juvenile survival reduced? Are the growth rates of adults depressed to the extent that the energy devoted to gametes is reduced or the time to reproduction is increased? Are there hormonal or other physiological changes that interfere with the reproductive cycle? Once we know why recruitment is reduced, it is critical to determine how the pollutant exerts its influence (what is its site of action, specific or general effect), at what levels it becomes noticeably toxic and what factors (intrinsic and extrinsic to the organism) influence its toxicity. It is these questions that toxicology can effectively address.

Studying the toxicological effects of pollutants on individual organisms (or their component parts) may sometimes allow generalizations to be made with regard to the effects of structurally or functionally similar chemicals, and may facilitate the formulation of simple predictions of impacts at higher levels of biological organization. As we will repeatedly argue, understanding and predicting the consequences of pollutant-induced effects on ecosystems requires that the effects be examined at the level of interest. Given the preponderance of reductionism in ecotoxicology, it is disturbing that we have very little idea about how often interactions among system components negate bottom-up predictions or to what extent higher level processes influence lower level phenomena as well. Most of us would look askance at a colleague who announced that he or she was setting out to discover the molecular mechanism of mercury poisoning by analysing the population dynamics of selected fish species. Would we find any less ludicrous the suggestion that studies of physiological response of isolated populations under laboratory conditions can be used to understand or predict the ecological consequences of pollutant exposure on ecosystems? Apparently so, as this is precisely what many ecotoxicological research projects and governmental regulations have been designed to do.

Is there enough experimental or theoretical evidence to justify the contention that physiological or suborganismal function can be related in a predictable way to ecosystem dynamics? Gross (1989) discussed the limitations of the reductionist approach, as applied in studies of plant physiology. He argued that this approach has led to substantial advances in our understanding of the details of plant physiology but may be of less use in studying large-scale phenomena. A bottom-up approach would attempt to understand an ecosystem as the biochemical and physiological sum of its parts. Although we would assume that we could omit some of the physiological details, we do not have a good feel for how much of the physiology can be safely ignored. As a case in point, Gross (1989) cited the study of the effects of increases in atmospheric carbon dioxide. He contended that 'although one can make an argument that the basic physiological questions are inherently interesting in their own right, completely different top-down methods may be much more appropriate ... A reductionist approach based on physiological details may be so incomplete that it is irrelevant, and it may even be counterproductive by taking attention and limited funding away from approaches more appropriate for the policy questions being posed.'

In contrast, Koehl (1989) argued that studying processes at the level of the individual organism can lead to greater understanding of ecosystem-level phenomena. For example, studies of individual organisms can reveal which factors can and cannot be ignored in the development of ecosystem models. Koehl (1989) cited a number of areas in which study at the interface between organismal and ecological levels of organization has led to valuable interactions between theory and data. For example, studies of biomechanics have led to a greater understanding of the physical constraints operating on organism function, which in turn can control ecological performance at the community level. To what extent a bottom-up approach can contribute to the solution of ecotoxicological problems will undoubtedly remain a subject for debate. We will return to this very important issue when we investigate the problem of validation of laboratory test methods (Chapter 6). Validation often is primarily concerned with extrapolating up the biological hierarchy from organism to population to ecosystem.

1.2 Pollution: what is it and what does it do?

The word pollutant, as defined by Moriarty (1983), refers to 'substances that occur in the environment at least in part as a result of man's activities, and which have a deleterious effect on living organisms.' Kinne (1968) defined pollution as 'human activities causing negative effects on health, resources, amenities or ecosystems', where negative effects imply demonstrable impairment. One of the more provocative, yet perhaps most accurate, definitions of pollution was presented by Bang (1980). He wrote, 'Pollution is largely an unashamed value judgment as to what one wishes to have or not have in the environment. The judgment may or may not be buttressed by scientific data, and nearly always involves a choice.'

Ecotoxicologists are primarily concerned with the fate and effects of chemical pollutants. Such pollution represents only one class of artificial interference in natural ecosystems. Others include the loss of habitat, the fragmentation of habitat (which produces deleterious area, edge, demographic and genetic effects), excessive noise, overexploitation, the spread of exotic species and diseases and climate change. Whereas the degree of chemical pollution tends to correlate with the level of industrialization, impacts caused by habitat destruction and extinction currently represent the primary human impacts in less-developed nations. Pollution arises from a large number and kind of human activities. Pollutant input into the environment can be from point (for example, industrial discharges) or nonpoint (such as agricultural or urban runoff, atmospheric deposition) sources and can occur via air, land or water with substantial cycling among these compartments.

1.2.1 Pollution of the coastal ocean

Approximately 70% of the human population now lives within 60 km of the coastal ocean, and this percentage is increasing (Gray, 1991). The stress experienced by coastal environments is clearly evident from the fact that of the 50 largest cities of the world, only seven are isolated from the coast. Half are situated directly on an estuary, with another third located further up river from an estuary (Squires, 1983). Estuaries have been considered by industry to be economically valuable sites for development given the possibilities for trade, traffic and ease of pollutant discharge. However, from an ecological perspective, estuarine environments are problematic because of their pronounced tendency to trap and accumulate particle-bound pollutants such as heavy metals and many organic chemicals (Bryan, 1980; Kinne, 1980).

Given the vastness of the world ocean, the concentration of human populations along coastal areas and large bodies of fresh water and the heavy dependence of industrial and agricultural activities on a ready water supply, it is not surprising that pollution studies have focused more heavily on freshwater and marine systems than on their terrestrial and atmospheric counterparts. This should not imply that pollution of land and air are necessarily less severe or important. Rather, it suggests that our knowledge of pollutant effects on terrestrial and atmospheric systems is more limited. As concern increases over the problems of global pollution and climate change, we will require a more complete and integrated understanding of pollutant fate and effects among land, sea and atmosphere. In the meantime, much of our discussion will draw on data and insights gained through the study of aquatic ecosystems. As this is also the area with which the authors are most familiar, we alert the reader to our bias while emphasizing that most of the issues we will explore are of general relevance.

One of the most important complicating factors involved in the study and especially the management of marine pollution is the fact that vast areas of the world ocean are used by all, but belong to none. Conflicts of interest among different nations can greatly impede attempts to identify and limit pollutant impacts and highlight the importance of international communication and cooperation. With these goals in mind, the 14th European Marine Biology Symposium, held in Helgoland in 1979, was dedicated to the 'protection of life in the sea'. Presentations at this symposium focused attention on the following classes of pollutants: metals, radionuclides, oil and oil dispersants, pesticides, industrial and domestic wastes and a few additional sources arising from such activities as dam building, power-plant discharges and deep ocean mining. The general conclusion of the symposium was that pesticides and related organic compounds presented the greatest and most pressing danger to life in the sea.

These substances are largely foreign to the marine environment, are often specifically designed to kill or incapacitate living organisms and tend to accumulate and exert long-term effects.

Pesticides The transformation of pesticides and other organic compounds by various biotic and abiotic processes greatly complicates the study of their fate and effects. Despite the fact that efforts over the last decade have sought to develop pesticides that firstly are more specific in their action, secondly are less persistent in the environment, or thirdly can be used in smaller amounts, such substances continue to present a serious threat to terrestrial and aquatic ecosystems. Although modern pesticides are generally more rapidly degraded than those used in the past, they can be extremely toxic to target and nontarget species alike (Hayes and Laws, 1991). For example, it has been shown that repeated applications of even non-persistent pesticides can have cumulative effects on the biota (Bookhout and Monroe, 1977). Furthermore, some of the currently used insecticides, such as the pyrethroids, have been found to cause biological damage at concentrations an order of magnitude below the current limit of detection (Schimmel *et al.*, 1983).

Heavy metals Heavy metals also present serious problems for aquatic ecosystems and may be of particular concern in coastal areas, especially estuaries. Metals enter ecosystems via atmospheric fallout, rivers, outfalls and dumping. They are released as byproducts of industrial processes, in association with mining activities and from ships. Dredge spoils can also contain large amounts of metals, as well as other pollutants. For example, it is believed that dredge spoils from New York harbour constitute the largest single source of contamination by metals into the New York Bight (Squires, 1983). Mercury, cadmium and lead are generally considered to be the most dangerous to humans and ecosystems, and copper, zinc, silver and chromium can also pose significant dangers. Other metals may pose problems in certain limited areas.

Oil Pollution from oil has had severe and very obvious effects, the most obvious of which have occurred as the result of spills or tanker accidents. Such acute inputs of oil to the ocean, which may be limited in space and time, have been estimated to account for approximately 10–20% of the total load to the world oceans (Sasamura, 1981 cited in Jensen and Jørgensen, 1984, Table 1.1). The remaining 80–90% of the oil entering the oceans occurs via chronic input such as municipal and industrial waste water, river input, urban runoff, routine emissions from ships and atmospheric fallout. The evidence continues to accumulate that a number of components of petroleum and coal are quite widespread in occurrence and have measurable effects on aquatic organisms (Neff, 1985; Granby, 1987). Recent studies of the effects of oil-drilling activities on the benthic fauna of the North Sea detected significant effects on the biota at distances of up to 2–3 km from the drilling field (Gray *et al.*, 1990). In some marine basins that have supported drilling activities, sediment background levels of oil appear to be increasing. For example, background values of around 300 mg of oil per kilogram of sediment dry weight have recently been

Table 1.1 Estimated inputs of petroleum hydrocarbons to the world oceans

Source	% Total input 1973	1981
Transportation, chronic	30.0	30.5
Transportation, acute	4.9	11.6
Offshore activities	1.3	1.4
Refineries	3.3	39.5 (Refineries, Municipal waste water, Industrial waste water, Urban runoff, Rivers combined)
Municipal waste water	4.9	
Industrial waste water	4.9	
Urban runoff	4.9	
Rivers	26.2	
Atmosphere	9.8	8.5
Natural seeps	9.8	8.5

Data from Sasamura (1981; cited in Jensen and Jørgensen, 1984).

measured in the Shetland Basin of the North Sea and are believed to be related to the use of oil-based drilling muds in this area (B. Kruse, personal communication).

Oil inputs associated with offshore drilling activities, drill cuttings, production water and accidental spills are the most significant sources of oil to the marine environment. In the North Sea, for example, 29 700 tons of oil were released in 1988 as a direct result of drilling. Seventy-six per cent of this amount was attributed to drill cuttings. As oil fields become exhausted, offshore operators pump up increasing amounts of oil-containing production water. A 1990 *Interim Report on the Quality Status of the North Sea* demonstrated an upward trend in the amount of oil discharged with production water between 1981 and 1988, with an increase of 63% between 1985 and 1988.

Although air surveillance is having a preventive effect in some cases, illegal discharges from ships and offshore production platforms remain a significant source of oil to the marine environment (Granby, 1987). In recent years, the use and discharge of highly toxic chemicals by offshore drilling operations is becoming a matter of increasing concern (B. Kruse, personal communication). Dispersants that have been used to clean-up oil spills present their own environmental problems. They may retard the evaporation of volatile components, they pose a potential toxic threat to biota, and their use tends to be costly and restricted to a local scale (Benyon and Cowell, 1974).

Domestic and industrial waste One of the major effects of domestic and industrial pollutants appears to be an alteration in nutrient dynamics of rivers and coastal areas (Pearson, 1980). Although the contaminated water itself may often pose little direct toxic threat, the organic load can have a substantial impact on aquatic environments by causing oxygen depletion. Provided that the impact is limited in area and duration, the effects on organism abundance and diversity appear to be relatively short-lived and reversible (Pearson and Rosenberg, 1978). However, in many freshwater and coastal areas, eutrophication is becoming a significant and chronic problem. Nutrient enrichment often occurs in combination with other, more toxic pollutants and can greatly exacerbate pollutant toxicity through its influence on oxidation-reduction chemistry (which can alter pollutant bioavailability) and because it exerts an added (though not necessarily additive) stress on the biota.

Litter Pollution in the form of litter, such as plastics, styrofoam and other substances that accumulate on beaches and which are often seen floating in coastal waters, has been considered as primarily an aesthetic problem. Although aesthetics are certainly a major aspect of such pollution, there may be a number of additional ecological effects. These have recently been summarized by Ross *et al.* (1991). Among the biological effects of litter are entanglement, 'ghost-fishing' and ingestion of plastics by marine organisms. Studies performed in the 1970s noted the widespread occurrence of plastic particles in coastal waters as well as in the Sargasso Sea, and small pieces of plastic were detected within the digestive system of several coastal fish species (Carpenter and Smith, 1972; Carpenter *et al.*, 1972). At a 1991 workshop held on San Miguel, Azores, Dr H. Martins from the Department of Oceanography, University of the Azores showed graphic evidence that the ingestion of large amounts of plastics by sea turtles is a significant cause of death in these endangered animals. Apparently the clear plastic is mistaken for medusae, which constitute an important food source for the sea turtles.

Economic effects of litter pollution include lost fishing opportunities, damaged market potential for fish and fish products, beach clean-ups, lost tourism opportunities and cost and time involved in repairing vessels and gear. Ross *et al.* (1991) reported on a 1989 survey of the types and sources of beach litter collected from Halifax harbour, Nova Scotia, Canada. They found that 54% of the litter was plastic, 12% was styrofoam, 12% was metal, 8% was glass, 5% was paper, 5% was wood and 3% was rubber. They determined that the largest sources of beach litter were recreational activities (32%) and land-based sources (30%). Sewage outfalls contributed 17%; 11% was from industry; 8% was from fishing; 0.8% was from shipping and 0.2% was from military sources.

To date, environmental scientists have identified a great number and variety of pollutants in natural environments around the globe. However, many potential pollutants are currently unknown either because we have not looked for them or because our analytical techniques are not sensitive enough or sophisticated enough to detect their presence at natural concentrations.

1.2.2 Direct and indirect effects

As discussed above, toxicologists are primarily interested in determining how chemical pollutants and other toxic substances impair function at the individual organism level. Toxicological studies concerned with the mechanisms of toxicant action often focus at the suborganismal level, but the individual organism is still the target of primary interest. For ecotoxicologists, impairment of an individual organism is of less significance, because the targets of primary concern are populations, communities and ecosystems. Bayne (1980) argued that, with regard to individual organisms, a pollutant-related (or other stress-related) change in a physiological rate process is only of significance if it can be shown that such change alters the individual's fitness. From an ecotoxicological perspective, pollutants are primarily judged on the basis of their interference with population dynamics and consequently for their significance as agents of selection and determinants of ecosystem structure and function. Blanck *et al.* (1988) argued that pollutants which do not exert a selection pressure cannot cause any significant biological disturbance in an ecosystem.

In general, the toxicological effects of a pollutant refer to lethality or to physiological, morphological or behavioural effects that interfere directly with organism function. Identifying the ecotoxicological effects of a pollutant is considerably more complex. This is because, in nature, pollutant effects can be mediated by physical and chemical factors in the environment which can act to enhance or inhibit the direct toxic impact on organisms. Pollutants can also have effects as a result of alteration of the physical habitat or effects on competitors, predators, symbionts, etc. For example, nutrient enrichment, which is a serious problem in many freshwater and estuarine habitats, causes damage primarily through effects such as hypoxia and changes in species composition. Another effect of pollution may be an increase in the susceptibility of organisms to bacterial or viral attack (Bang, 1980; Frederick and Pilsucki, 1991). Many of the effects of pollutants can be difficult to predict as they often involve complex and possibly nonlinear interactions among biological as well as chemical and physical components of the system.

Moriarty (1983) categorized pollutants into two main classes, those having **direct** toxic effects and those having **indirect** effects through alteration of the physical or chemical environment. In other words, some pollutants may disrupt physiological (or molecular or biochemical) structure and function and others may primarily disrupt ecological structure and function. The former are often referred to as toxins or toxicants. [According to Hayes (1991), **toxicant** is a general term for a poisonous substance whereas **toxin** originated in reference to specific substances of plant, animal or bacterial origin that were labile, proteinaceous and capable of giving rise to specific tolerance in humans and other mammals. Through common use, the term toxin has eroded into a synonym for poison or toxicant. We specifically use toxin to refer to those toxicants produced by living organisms.] Toxic effects may appear immediately or only after accumulating above a certain level in cells or tissues.

According to Moriarty (1983), toxicants (direct-effect pollutants) can be distinguished from substances that disrupt ecological function, such as predator–prey relationships, competition, trophic relationships, etc. (indirect-effect pollutants). However, in practice, the distinction between pollutants having direct and indirect effects is often far from clear. For example, a pollutant that decreased prey abundance might increase intraspecific competition in a predator species. We would define this as a direct effect on the prey species and an indirect or ecological effect of the pollutant on the predator species. Intertidal gastropods respond to oil emulsifiers, and probably many other toxicants, by retracting into their shells (Baker and Crapp, 1974). Such behaviour is a sublethal response which presumably helps these organisms to limit their contact with the contaminant, that is, withdrawal is a direct behavioural effect of the emulsifier. However, such behaviour results in snails being swept away from the shore where they may be more vulnerable to predation or be unable to return to a suitable habitat (indirect effects of emulsifier exposure).

As many pollutants can be expected to exhibit both direct (toxic) and indirect (ecological) effects, we suggest that it may be more useful to classify the effects instead of the pollutants. Thus we extend Moriarty's basic definition of ecotoxicology to the following: **the field of study which integrates the ecological and toxicological effects of chemical pollutants on populations, communities and ecosystems with the fate (transport, transformation and breakdown) of such pollutants in the environment.**

1.2.3 Fundamental similarity of pollutants to natural disturbances

Sometimes pollution occurs via catastrophic events of relatively short duration, such as the *Exxon Valdez* oil spill in Alaska. In such cases pollutants can have very rapid and severe impacts on ecosystems, in the same way that storms, volcanic eruptions and fire can effect massive environmental change. In other cases, pollution occurs via chronic low-level disturbance such as the continuous input of chemicals via wastewater or release of volatile substances to the atmosphere through combustion of fossil fuels. Such perturbations are analogous to, for example, environmental changes resulting from local shifts in rainfall level or temperature. Chronic perturbations are more difficult to distinguish spatially or temporally and may extend for long periods of time relative to the generation time of community inhabitants.

The reason that such analogies can be made between pollutants and naturally occurring perturbations is that the response of living systems to both types of stress is, in general terms, quite similar. Independent of whether stress is imposed on an ecosystem from natural or anthropogenic factors, the biological response occurs via individual organism acclimation, followed, at increasing levels of stress, by selective elimination of less tolerant genotypes within populations and selective reduction or elimination of less tolerant species. Both natural environmental disturbances and anthropogenic perturbations can present completely novel challenges to biological systems or can impose stresses purely from quantitative changes that are outside the 'normal' environmental range. The duration of both natural and artificial perturbations can be either temporary or prolonged. In both cases, the effects can be acute or chronic, or both, and may require some threshold level to be reached before detrimental effects appear.

Xenobiotic substances (as defined by Rand and Petrocelli, 1985) have often been set apart as fundamentally unique because they are entirely artificial and therefore new to nature. But we contend that many natural perturbations or occurrences are just as unique, from an organism's or species' perspective. New and highly toxic chemicals have been made by organisms in the past. In fact, some of the most toxic substances known are those of natural origin (Ray, 1991). For many naturally occurring toxins, we have no more reason to expect that an organism has the necessary machinery with which to respond than it would to a artificial toxicant. When resistance* of pest species to insecticides first appeared there was uncertainty about it because it seemed unlikely that a natural population could carry genes for resistance against synthetic toxicants to which it had never been exposed. It has been shown very clearly that resistance arose through selection acting upon natural variation already present in the population before pesticide exposure (Wood, 1981). Through the course of evolution, organisms have repeatedly had to adjust to the addition of new 'natural' chemicals. Thus, it is significant that the founders of ecotoxicology explicitly included 'naturally existing pollutants' as within the scope of ecotoxicology (Truhaut, 1977).

More often than not, biological systems are faced with a mixture of stresses, of natural or anthropogenic origin or both. The responses that ensue occur regardless of whether the environmental perturbation is due to 'natural' or 'artificial' changes. Any differences that may occur between the responses of living systems to natural versus artificial pollutants are generally more constructively viewed as quantitative, rather than qualitative, differences. At the risk of flogging a dead horse, we emphasize that from the ecosystem's point of view, there is no fundamental difference between an ecological impact arising from a natural chemical and an ecological impact arising from an artificial chemical.

1.3 What is the scope of ecotoxicology?

We have been struck by the difference of opinion among scientists regarding the scope of ecotoxicology and to what extent it is a unique and separate discipline, one that should be treated as distinct from ecology and toxicology. It is our intention here to attempt to clarify the scope of ecotoxicology. First, does ecotoxicology qualify as a separate field? Is it, as Pratt (1991) suggested, a set with no members? To answer these questions, we need to determine whether we can identify fundamental principles which can be viewed as unique to a discipline of ecotoxicology. These may require the use of new or characteristic

*Although some authors distinguish between the terms resistance and tolerance, we follow Weis and Weis (1989) and others and use these two terms interchangeably.

methodologies, though it should be noted that new methodologies (while perhaps facilitating the discovery of new principles) are not sufficient for defining a new field. Second, and perhaps more important, we have to ask what ecotoxicology would stand to gain by distinguishing itself as distinct from ecology and toxicology.

The following observations have been derived mainly from formal and informal conversations with ecotoxicologists and scientists in related disciplines from different countries in Europe and North America. Our intention in the following section is not to aggravate or alienate, but to attempt to understand firstly, why the objectives of ecotoxicologists in academia, industry and government often seem to be at odds, and secondly, how this situation can be rectified. Most work involving ecotoxicologists is designed to meet one or more of the following objectives (not in order of importance).

1. Generate data that can be used to support decisions for risk assessment and environmental management.
2. Meet legal requirements regulating the development, manufacture or release of potentially dangerous substances.
3. Develop empirical or theoretical principles to further understanding of the behaviour and effects of chemicals in living systems.

Thus ecotoxicological findings provide an essential part of the necessary data for decisions involved in risk assessment and environmental management. In principle, ecotoxicological research should influence the development of regulatory guidelines for environmental protection, and the design and performance of routine testing and consultancy. Much of the routine testing is performed in order to satisfy regulatory requirements (for use in chemical classification and labelling schemes, for example) and to provide data for use in risk analyses. It is unfortunate that, although such data could potentially contribute a great deal toward the study of ecotoxicological phenomena, much of it remains hidden in company and government files.

Within the realm of scientific research in ecotoxicology, we came upon some rather surprising distinctions. Many of the ecotoxicologists with whom we spoke clearly distinguished their work as either basic or applied ecotoxicology. In a rather confusing chairman's summary from a session of the First European Conference on Ecotoxicology, Persoone and Kihlström (1988) subdivided European research into the testing of ecotoxicological effects into three categories. They defined **applied** ecotoxicological research as those studies directly involved in the development of standardized tests for regulatory purposes. Contrary to standard usage (in which the terms 'basic' and 'fundamental' are used synonymously), they distinguished **basic** dose–effect research from **fundamental** ecotoxicological research. The former they characterized as an amassing of empirical dose–effect data at various levels of biological organization. To the latter category were relegated **classical** toxicology (such as physiological mechanisms and kinetics of toxicity) and **supraorganismal** ecological toxicology. It is difficult to fathom the usefulness of such distinctions, either in theory or practice. If this is an accurate description of the way in which ecotoxicology is practised, it serves to explain much of the confusion of goals and lack of focus that frustrate many ecotoxicologists.

During discussions with practising ecotoxicologists, the opinion was expressed to us that applied research in ecotoxicology involves the investigation of problems within particular strategic areas. In practice, strategic refers to work that is directly related to the development of test protocols or risk assessment schemes. In these strictly applied studies, gaps in existing data are identified and specifically addressed. Those who defined themselves as working in applied ecotoxicology defined basic research in ecotoxicology to include research into artificial substances or other toxicants which had no application to 'real world' problems, either because the substances used for study were not of great environmental concern or because the concentrations used were completely unrealistic. In addition, the topics considered in basic studies are often considered of little relevance for regulatory or risk assessment purposes because such studies focus on understanding scientific details. Whereas such details may be absolutely critical to increase scientific understanding, they are often considered to be of low priority by environmental managers because they may not significantly or immediately alter management decisions. The casual dismissal, and sometimes complete lack of interest, exhibited by regulators in response to new research findings is inexpressibly frustrating to scientists and, we suspect, may discourage many good scientists from pursuing research in ecotoxicology.

At a recent conference, we were surprised to overhear a colleague complaining about 'all of these academics, muddling about, with no conception of the relevant problems'. Interestingly this view is inconsistent with the fact that ecotoxicologists working in academia have often been among those to raise the loudest voice against the ecological irrelevance of current testing procedures. This voice of reason is countered by an unfortunate tendency among some academic scientists to assume that any research going on outside of the university is second (or third) rate. We clearly have a serious communication problem. If academia does, in fact, ignore the call for theoretical or empirical study of particularly relevant problems, then their lack of responsibility is justly criticized. On the other hand, if academia is largely unresponsive to many of the concerns and needs of ecotoxicologists in government and the private sector, it may be a direct result of the fact that much of the information is buried in a 'grey literature' which is circulated only among a limited clique. Whether this is any great loss is questionable, given that the absence of independent peer review is a generally reliable indicator of reduced scientific quality.

There are two fundamental problems in attempting to distinguish scientific research into 'applied' and 'pure' categories. The first is a problem that is common to all areas of scientific research. It is essentially the fact that any differences between so-called pure and applied research are more apparent than real. In a 1991 editorial in *Science*, Koshland summarized the problem.

> The line between pure research and practical application has always been difficult to draw. Some like to claim they are doing pure research with no practical purpose in mind. There is an implication that "pure" is not only nobler but also more difficult intellectually than "applied" research. Equally vociferous are those on the applied side who suggest that they labor for the good of mankind, whereas the ivory tower types are simply enjoying themselves. Those demarcations are gone forever, or should be. It is a wise person who can state with certainty what basic research result will never be practical, or that applied research will not lead to new basic insights, or which is intellectually more demanding.

Are there any among us who can claim to be so wise? In an introduction to the widely used textbook, *Ecology*, Ricklefs (1979) addressed the question of how we should expend our research efforts in ecology. Should we focus on the solution to immediate problems or on expanding our understanding of the natural world? 'To forsake the intense pursuit of ecological theory merely because our immediate problems require the application of knowledge would be a grave error of shortsightedness. Experience has shown that solutions to problems are often extracted from or based on knowledge originally acquired in studies that had no obvious application. So, too, have many advances in theoretical ecology come from applied research.'

We have not been able to discover how basic and applied research have come to be regarded as mutually exclusive activities. Although many would agree with Ricklef (1979) and Koshland (1991) – that it is impossible to foresee how so-called 'basic' and 'applied' research will benefit each other and that the division of science into such categories can be destructive – the fact is that these divisions remain. We suggest that their elimination could go a long way towards improving communication among ecotoxicologists and could enhance the development of research programmes that are of both scientific and practical significance. Particularly in a field such as ecotoxicology, which specifically addresses problems related to human impact on the environment and the ways in which those environmental impacts reflect back on us, can the term 'pure' research have any useful meaning?

The second problem with attempting to distinguish 'applied' from 'pure' research is specific to ecotoxicology. We have discovered that what is often meant by applied ecotoxicology hardly qualifies as research at all. The term, 'applied' often appears to be reserved for the performance of old tests on 'new' chemicals or species, or perhaps for the design of a 'better' shaker-flask. Such efforts not only perpetuate the use of recognizably inferior test systems, but they divert attention and funding away from developing approaches that will truly improve our ability to successfully understand and regulate chemicals in the environment.

It has been our experience that some of the most badly needed research in ecotoxicology goes unfunded for no other reason than that it does not easily fit into the artificial categories of 'basic' or 'applied'. Thus, a proposal addressing the genetic adaptations of populations to pollutant stress is likely to be turned down by the natural sciences division of a government funding agency as being 'too applied' and simultaneously criticized by the environmental protection agency as 'too basic'. The primary strength of this project – that it contributes to our understanding of important biological phenomena (that is, evolution) while also providing pertinent information regarding the effects of pollutants on natural populations – becomes its weakness. Would it not be more effective to expend less energy on categorizing our research activities and more energy on ensuring their scientific rigour?

In their plenary address at the Second European Conference on Ecotoxicology, Cairns and Pratt (1992) referred to the needs of the **ecotoxicology industry**. This perceptive description of ecotoxicology as an industry reflects the fact that much of what passes for science in ecotoxicology is actually little more than the promotion of products (that is, tests, guidelines etc.) by self-serving interest groups whose concerns appear to be limited more by economics and politics than by science. This is not to say that there are not many concerned and dedicated scientists who have and are contributing to the science of ecotoxicology. However, except for a few outspoken and articulate critics, the concerned voices of many scientists appear to be largely ignored.

1.3.1 Relationship of ecotoxicology to other disciplines

Politics aside, we are still left with the question as to whether, and for what purpose, ecotoxicology should stand apart as a separate discipline for study. We have argued that, from an organism's or ecosystem's point of view, there is no fundamental difference between perturbations due to natural stresses and those due to artificial pollutants. So what makes ecotoxicology different? Is it more or less than the sum of its ecological and toxicological parts? In Figure 1.1, we have attempted to locate ecotoxicology in relation to the various disciplines to which it is most closely related, and in the following lines we describe the primary focus and relevance of each of these disciplines to ecotoxicology.

Ecology is the study of the structure and function of ecosystems. It is a field composed of many subdisciplines and one which often focuses on the study of interactions among ecosystem components, both living and non-living. The single most important principle in ecology, which gives pattern and meaning to the diversity of life, is that of evolution (Ricklefs, 1979). Ecology is concerned with the pathways and rates of transfer of energy and material within and between ecosystem components (fate) and with the effects on ecosystem components (structure and function). Ecologists study processes occurring at the organism, population, community and ecosystem levels of organization. Neuhold (1986) made the interesting suggestion that, in some ways, toxicology

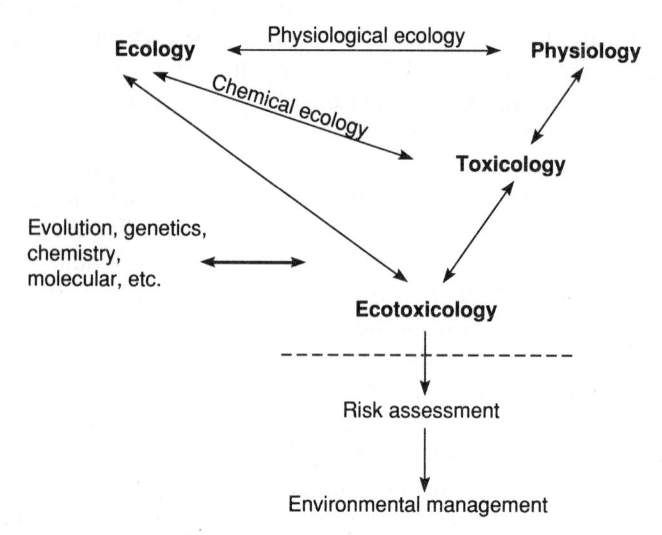

Figure 1.1 The relationship of ecotoxicology to other scientific disciplines and to environmental risk assessment and management. Whereas ecology and physiology generally deal with the study of 'normal' systems, toxicology and ecotoxicology deal with damaged systems. Ecotoxicological research provides the basis for strategies of risk assessment and for environmental decision making.

can be considered as a subset of ecology; one that is concerned specifically with the study of the individual in relation to its environment.

Physiology is the study of the structure and function of organisms. Two primary objectives of modern physiology are to understand the central unifying principles by which physiological function is governed and to understand the physiological diversity in biological molecules, cells, tissues, organs and organisms (Feder *et al.*, 1987). As discussed earlier, toxicology is concerned with identifying and quantifying the injurious effects of physical and chemical agents on organism structure and function. Whereas ecology and physiology are most often concerned with understanding the structure and function of systems under 'normal' conditions, toxicology and ecotoxicology study the structure and function of perturbed systems.

Not only can ecotoxicology draw on principles from ecology and toxicology, it can provide new insights into these fields as well. Within ecology, divisions into various subdisciplines such as evolutionary ecology, population ecology, community ecology, systems ecology etc. have been criticized as leading to a lack of communication and synthesis among subdisciplines and to limitations in training and vision for the field as a whole (Tilman, 1989). Ecotoxicologists could benefit from this discussion and should seek to enhance rather than sever their connections with related disciplines and subdisciplines. Provided that the place of ecotoxicology and its connections to other areas of study are recognized and remembered, then whether or not we consider ecotoxicology to be a separate field is really of little importance. If defining ecotoxicology as a separate discipline helps to focus research questions, facilitates the administration of research grants, and aids academic departments in achieving a balance of study areas, then it should be encouraged. If, however, it serves to isolate ecotoxicology from its scientific roots, it will lead to little more than re-inventing the wheel and should be strongly discouraged.

Today, many ecotoxicologists appear to be unaware of the early contributions in ecology as well as many recent and relevant accomplishments in the study of ecosystems. This is undoubtedly a direct consequence of the fact that ecotoxicology lacks any ecological roots, having been 'created' by toxicologists. Moriarty (1983) attempted to bring a more ecological perspective to the study of ecotoxicology, but interestingly, his view of ecotoxicology is criticized by many ecotoxicologists. Cairns and Mount (1990) lamented, 'It is a pity that noted ecologists and ecotoxicologists interact only infrequently.' Ecotoxicologists are also in danger of cutting themselves off from certain developments in toxicology, particularly that area related to the study of natural toxins. Antiherbivore, antifungal and antibiotic substances produced by living organisms can provide important insights into how organisms respond to toxicants (physiologically and ecologically) and can lead to the development of more sophisticated means of pest control.

We suggest that neither Truhaut's (1977) classification of ecotoxicology as a subdiscipline of toxicology nor Moriarty's (1983) classification of ecotoxicology as a subdiscipline of ecology are entirely accurate. Ecotoxicology draws on principles from ecology, physiology, chemistry and toxicology, as well as from a number of other areas such as biochemistry and molecular biology. Whereas information from toxicological studies often finds application in medicine, ecotoxicological research has applications for the analysis and control of toxic substances in the environment. The theoretical developments and empirical findings of ecotoxicological studies are used directly to assess the risks of artificial substances and other human activities for the environment. The estimated risks are considered, along with political, economic and cultural concerns, in the development of management strategies. Thus, management decisions are based on risk assessment. The criteria used for risk assessment are strongly influenced by the needs of management. Ideally, a feedback system between ecotoxicological research, risk assessment procedures and management needs would assure that decisions are based on rational, scientifically sound information and that research efforts are directed toward questions of greatest social, economic and ecological concern.

As public concern mounts over the effects on the environment of changes made by humans, the need for integrating scientific knowledge with political, economic and cultural issues is increasingly apparent. How does ecotoxicology fit into this picture? As a science, ecotoxicology is a means whereby understanding of phenomena is increased through the development of testable hypotheses. In this case, the hypotheses deal with the study of the environmental effects of chemical pollution. One aspect of this study concerns the formal assessment of the potential risks or hazards arising from human activities. In turn, risk assessment provides a rational basis for management decisions. Ideally, environmental management incorporates scientific information with political, economic and cultural concerns.

1.4 How are assessment and prediction accomplished?

1.4.1 Testing in ecotoxicology

Broadly speaking, as ecotoxicologists our primary interest is in assessing and predicting the effects of pollutants. How do we go about this task? Just as there are many different biological levels at which pollutants can exert their effects, there are a variety of methods available for investigating the type or extent of an effect. Our choice of method must depend on the specific question that we address. Virtually all of the tests used by ecotoxicologists are designed to determine the fate or behaviour of pollutants, the biological effects of pollutants, or both. Tests can be classified according to design, level of biological organization, exposure period and endpoint.

Tests used in ecotoxicological studies are **designed** to be performed either in the laboratory, in the field or by computer. Laboratory tests provide a greater degree of control but generally deviate further from the actual system than do field tests. Historically, most of the tests designed to address ecotoxicological questions have been performed in the laboratory. Risk evaluations, used in environmental management, are made primarily on the basis of laboratory results. However, field validation is often considered as an essential ingredient in the risk evaluation process.

Various theoretical and analytical models have become important ecotoxicological tools for the study of pollutant-related phenomena. Most models developed to address ecotoxicological problems have been of the descriptive (that is, predictive) variety (for example, Quantitative Structure Activity Relationships (QSARs), acute-to-chronic toxicity relationships), and there is an obvious need for greater development of analytical (theoretical) models that can provide a conceptual framework for focusing the future direction of ecotoxicological research. For example, life history models can be used to predict a population's optimal response strategy under different degrees of growth and mortality stress (Sibly and Calow, 1989). Predictions for model systems based on resource competition theory have revealed that the availability of resources, often overlooked in ecotoxicological studies, can have a crucial effect on the impact of a toxicant on a community (Landis, 1986). Computer simulations have been used to provide useful generalizations about complex systems (Wilson and Botkin 1990, for example) and may provide an important step in risk assessment. The exact approach we take toward model building will depend on the purpose we have in mind; whether it is an attempt to understand the key features of a particular system or 'simply' to make predictions. A model can be of little use as a predictor but of great value as a theory, and vice versa. Descriptive models implicitly assume that conditions do not change and that the model parameters adequately represent the causal pattern of interest (Koehl, 1989). They are therefore often limited to short-term or very narrowly defined predictions. Because the interactions among the components are based on empirical correlations and relationships that are not well understood, it is difficult to extrapolate to conditions with different parameter values.

A second means of classifying tests relates to the **level of biological organization** that is studied. Single species, multispecies and ecosystem level tests offer different insights into the behaviour and effects of pollutants. Single-species test systems are most commonly employed in studies measuring toxicity, largely as a result of their historical precedence in the field of toxicology and because they are generally simpler to perform and interpret than higher-level tests. However, as Rand and Petrocelli (1985), among many others, have noted, 'the results [of single species tests] cannot be used to assess the chemical impact above this level of biological organization.'

Multispecies, community and ecosystem level tests permit the detection of indirect pollutant effects that are the result of species interactions, as well as direct toxic effects. Multispecies tests may use two or more species within a single trophic level (for example, Cairns *et al.*, 1986) or may combine species from different trophic levels in an artificial food chain (e.g., Kersting, 1984, 1985). In contrast, community and ecosystem level tests attempt to simulate critical processes occurring in intact ecosystems or parts of ecosystems (Davies and Gamble, 1979). Community and ecosystem test designs have included small laboratory bench-top microcosms (we do not distinguish between microcosms and mesocosms), large land-based tanks (such as Marine Ecosystems Research Laboratory (MERL), located at the University of Rhode Island, USA) and various sizes and designs of enclosures *in situ* (for example Davies and Gamble, 1979). Although the difference between a community-level test and an ecosystem-level test may not be obvious, we suggest that a useful distinction may be that the former term be used for those tests concerned primarily with biological effects (such as Wängberg *et al.*, 1991) whereas the latter term should be used for tests in which both pollutant fate and effects are measured (such as Controlled Environment Pollution Experiment (CEPEX) described by Menzel, 1977).

One of the primary types of test classification is based on the **length of the exposure period**. Tests are distinguished as acute, subchronic or chronic. The length of the exposure period is defined relative to the length of the test organisms' life cycle (Rand and Petrocelli, 1985). For most vertebrates and invertebrates, acute tests are generally performed for periods up to a maximum of 96 hours. Subchronic exposures typically extend for less than an entire reproductive period, whereas chronic exposures may extend for one or more life cycles.

A fourth criterion for test classification relates to the measured **endpoints**. Both lethal and sublethal endpoints have been investigated in ecotoxicological studies. Tests are often designed to measure the exposure concentration at which 50% of the test population is killed (LC_{50}) or exhibits a defined sublethal effect (EC_{50}). Sublethal endpoints include growth, reproduction and behaviour as well as various physiological, morphological, biochemical and molecular indices. Opinions differ as to the reliability of these various endpoints. Although lethality often provides the first measure of toxicity for screening of chemicals, it is generally believed that death is a crude endpoint and that various sublethal measures should be included in any evaluation of chemical toxicity.

Although the tests described above may differ substantially in terms of their inherent assumptions, technical requirements and the scientific expertise required, they share the common goal of estimating the effects of potentially harmful substances on living systems. Thus, we define an **ecotoxicity test** as any experimental measure (observational or manipulative) designed to assess the fate or effects of chemical pollutants at (or across) any level(s) of biological organization. This definition perhaps differs from common usage by its inclusion of environmental monitoring studies and physicochemical 'tests' of substances under the heading of ecotoxicity test. We include these because they suit the general purpose of ecotoxicity tests at least as well as those approaches included in standard usage of the term. However, we do make the distinction between **observational** tests and those that are **manipulative** or **experimental** in nature. The former would include field monitoring, measurements of the physical or chemical properties of a substance, observations of physiological or biochemical markers in polluted versus unpolluted habitats or along a pollutant gradient. The latter would include tests that measure response to experimental manipulations. As will become obvious, the development of ecotoxicity test methods has not proceeded equally at all levels. The formulation of tests to study pollutant effects at some levels (the whole-organism or physiological, for example) has been relatively straightforward, and such tests have been widely implemented. However, as we will discuss in some detail (Chapter 4), the usefulness of such tests is limited. Development of tests that directly measure pollutant effects on communities or ecosystems has proved to be more challenging given the high degree of complexity of these systems.

1.5 Who performs ecotoxicity tests?

While coherent strategies for assessing the potential hazard of those chemicals already existing in the environment are currently lacking, present regulations in many countries stipulate that the primary responsibility for the testing of new substances lies with those producing or releasing them (that is, industry). Often this is accomplished with the aid of private testing firms who specialize in performing a range of chemical and biological tests, under contract, as required by governmental legislation. Ensuring that legislated guidelines are followed, and assessing (and reassessing) the appropriateness of requirements concerning the testing or release of potential pollutants is carried out mainly by government regulatory agencies. Investigation of pollutant effects (or potential effects) that go beyond the realm of those required by law has been largely the task of academic ecotoxicologists. Are the goals of these different groups at odds? In theory, no, but in practice they may well be. In a very general sense, ecotoxicologists in every branch of study are concerned with determining the fate and effects of chemicals in nature. Presumably, protection of human health and environmental integrity are shared goals. However, the priorities and biases of government, academia and industry differ enough that we often appear to be more at odds with each other than in accord.

Given the importance and complexity of the issues at stake, it would seem critical that scientists and administrators involved in different branches of ecotoxicology pool their efforts to keep pace most effectively with an ever increasing demand for information. There are a number of ways in which such a division of labour can be facilitated. One practical objective, upon which ecotoxicologists in all branches of study agree, is that there is a great need for more extensive data on the toxic effects and behaviour of substances under natural conditions. As the largest burden for testing new chemicals currently falls on industry and consulting firms, ecotoxicologists working in this branch are in a position to accumulate huge amounts of data on the physicochemical and biological properties of a wide range of substances. Such data are vital in assessing current test procedures and in developing improved test systems. But unless this data is made readily and widely available, the efforts of those collecting it are largely wasted.

Ecotoxicologists working in the private sector (industry and consultancy) are paid to perform tests that will provide data for risk assessment and management decisions. Given that the primary objective of such tests is to fulfil legislative requirements, such as obtaining approval for the manufacture of a new chemical, greater emphasis is often placed on completing tests rather than on ensuring that they are the most relevant. A variety of incentives encourages reliance on available test methods, even though data provided by such tests may be inadequate in many respects. Ecotoxicologists working in academia have a different set of priorities and incentives. They are encouraged to challenge existing dogma and prevent a false sense of security in accepted approaches. Ecotoxicologists in academia approach this task with an almost missionary zeal and find great satisfaction in criticizing what they feel are oversimplified approaches to complex questions. But when asked by those in industry or government how to overcome the limitations inherent in current approaches, they rarely seem to give satisfactory answers. The scenario goes something like this,

ACADEMIC ECOTOXICOLOGIST: 'Single-species acute toxicity tests are too simplistic and have no connection with what is really going on out in nature. These standard tests are not only irrelevant and a waste of time, they may in fact do more harm than good if they lead us to believe that we can use them to adequately protect the environment when in fact we cannot.'

INDUSTRIAL ECOTOXICOLOGIST: 'These tests may be oversimplified, but they are also cost-efficient, easy to perform, the procedures have been worked out, and the fact is they are required by the government. We have absolutely no incentive to do more than is required by law, and, frankly, you have given us little hard evidence that current test procedures do fail to protect the environment adequately.'

GOVERNMENT ECOTOXICOLOGIST: 'Do you have any idea of the number of new chemicals that we have to assess each year? We can't tell industry to stop producing new chemicals and we can't wait until we understand the whole system before we try to protect it. If you think current procedures fail, then come up with some better tests – which must of course be simple, cheap and fast.'

ACADEMIC ECOTOXICOLOGIST: '(Pause) ... Well, it's very complex, and of course I'll need much more data before I can give you an answer. But those single-species acute tests are oversimplified and have no connection with what is really going on out in the field ...'

GOVERNMENT ECOTOXICOLOGIST: 'We need tests! Give us tests!'

Certainly, we can all sympathize with government ecotoxicologists' formidable tasks. Thousands of potentially hazardous chemicals are already present in our environment, with hundreds more added each year. Decisions involving the banning, limitation or free use of these chemicals falls largely on government ecotoxicologists. Theirs is not an enviable position. On the other hand, government regulators often appear unwilling to accept the costs of obtaining the necessary information to make a good decision, and they refuse to recognize that simple indices of pollutant danger or impact may not exist.

To some extent academic ecotoxicologists are doing their job in criticizing current approaches, but only half of their job. The other responsibility of ecotoxicologists working in the academic sector is to replace the existing dogma with new theories, that is, to provide a conceptual framework within which scientifically sound approaches to problem solving can be developed. If an industrial and academic ecotoxicologist were to name the attributes of the ideal toxicity test, these lists would probably be almost identical, however their rank order would likely be quite different. For example, 'low cost' would be closer to the top of the industrial ecotoxicologist's list whereas 'ecological relevance' would most likely head the list of the academic ecotoxicologist. Finding a suitable compromise of priorities can perhaps best be served by those ecotoxicologists working in government regulatory agencies. The role of such agencies – to make sure that the rules are appropriate and that they are followed – ideally includes both research into the effectiveness of current test methods as well as periodic review and updating of legislation.

To effectively fulfill this role, government ecotoxicologists must be aware of the needs of industry and must heed the cautionary warnings of researchers. Selecting testing criteria that balance the requirements of industry and science demands an accurate assessment of the true costs and benefits of artificial substances to society. Because, in practice, the government regulatory agencies are in the strongest position to effect changes in our approach to environmental problems, it is clearly to the benefit of ecotoxicologists working in other branches of the field to assist these agencies in making informed decisions. As we will continue to stress, one of the biggest obstacles preventing integration among ecotoxicologists in different branches of study is a lack of communication.

1.6 Summary and conclusions

Ecotoxicology can be defined as the field of study that integrates the ecological and toxicological effects of chemical pollutants on populations, communities and ecosystems with the fate (transport, transformation and breakdown) of such pollutants in the environment. Among the many pollutants that are recognized threats to human health and the environment are various organic chemicals and pesticides, heavy metals, oil (and various substances associated with its extraction), excess nutrients and various types of plastics. Ecotoxicologists, working in government, industry and academia, have developed a wide range of test systems for estimating the fate and effects of pollutants on living systems. Ecotoxicity tests can be classified with regard to design (field, laboratory, computer), level of biological organization (population, assemblage/community, ecosystem), exposure period (acute, subchronic, chronic) and endpoint (lethal, sublethal).

Ecotoxicology officially originated in 1969, as an offshoot of toxicology, largely in response to increasing public concern about the dangers of chemical pollution to human health and the environment. Although the term 'ecotoxicology' implies a linking between ecology and toxicology, the historical development of the field has been characterized by a strong bias toward toxicology. We believe that future development of the field will be enhanced dramatically if ecotoxicologists more fully exploit a large and relevant literature in the fields of ecology and evolution, particularly that body of work treating biotic responses to stress.

2 *Decision making in ecotoxicology: science and society meet*

Within a few hundred years this planet will have
little more than lineages of domestic weeds, flies,
cockroaches, and starlings, evolving to fill a
converted and mostly desertified environment left
in the wake of nonenvironmentally adaptive
human cultural evolution.
Terry L. Erwin, 1991

Given that growing public and political awareness of environmental contamination led to the origin of ecotoxicology, it is not surprising that the direction and scope of ecotoxicology, more than many other scientific disciplines, is influenced by nonscientific concerns. In this chapter we begin by examining the roles of the public and mass media in shaping environmental issues. We explore the criteria that enable us to decide which environmental components our efforts should be designed to protect and which changes arising from anthropogenic activities we deem unacceptable. We suggest that the term **ecosystem health** be replaced by other terms that more clearly underscore the important role of value judgments and desirability in the design of environmental protection and management strategies. In particular, we support the argument that conservation of biotic and abiotic diversity is a feature that should receive top priority. The last part of the chapter is devoted to a general discussion of risk assessment and risk management. We consider how these practices make use of ecotoxicological information, we discuss their scope and limitations, and we provide examples of how the assessment and management of risk are carried out in practice.

2.1 Scientific and public perceptions

Pollutants can have drastic effects on organisms or ecosystems that may be very subtle to the unpracticed eye (or that are not even visible without sophisticated and expensive scientific equipment). By the time that enough damage has occurred to be obvious to casual inspection by the average citizen, the situation may have already reached the stage of ecological disaster. On the other hand, some environmental problems receive much more public attention than is warranted based on the real risks involved because of political, historical or cultural factors or as a result of misinformation. Thus, society is largely dependent on scientists for providing information relevant to society's continued well-being, whereas scientists are charged with interpreting relevant scientific findings in a language that the nonspecialist can understand. This is quite a challenge, as our current level of understanding of ecosystems is such that we often have trouble interpreting the data in a way that is meaningful (or at least convincing) to other scientists. Recent debates over the effects of tropical deforestation on species extinction rates highlight this point (Mann, 1991). The best scientific estimates of the rates of species extinction vary by several orders of magnitude. Estimates vary due to uncertainty in assumptions relating to rates of habitat loss, the shape of the species–area curve and the absolute number of species. Although most scientists agree that tropical deforestation is leading to greatly increased rates of extinction, many feel that the lack of data diminishes the credibility of predictions and actually impedes efforts to preserve biodiversity.

Whereas gradual environmental degradation continues almost unnoticed, governments and the media often overreact to sudden events of lesser overall impact. This short-term mentality is reflected in political and economic policies that favour quick results and profits (Soulé, 1991). Our response arises, in part, from a misconception of risk. Because humans are evolutionarily programmed to respond to sudden dangers, we have difficulty perceiving small, gradual changes in our surroundings (Ornstein and Ehrlich, 1989).

Residents in and around New York City have found themselves in a state of panic on a number of occasions over the years. In 1973, unfounded fears that sewage sludge dumped into New York Bight was headed toward New York's recreational beaches resulted in massive media attention devoted to the 'Sludge Monster', created an economic loss for New York's tourist industry and caused a costly redirection of resources for federally funded research projects (Squires, 1983). Several years later, in the summer of 1976, an unusual series of accidents and atmospheric conditions led to the accumulation of a bizarre medley of debris on New York beaches.

> More informed citizens came to the conclusion that some threshold of environmental degradation had been reached and that the Bight had been "ruined". People stayed away from the beaches all summer in droves because of the perception of polluted waters and possible health

hazards, and because of the appearance of the beaches. And well they might have. What was the material which floated up on the beaches? It was a mix of natural and waste materials: seaweed, eelgrass and reedy materials; burned wood fragments; plastic debris including fragments of styrofoam cups, plastic bottles and beverage containers, tampon inserters, plastic sheeting fragments, straws, bottle caps, sanitary napkin liners, toys, cigarette filters and cigar tips, and so forth; condom rings and fragments; grease and tar balls of various sizes and appearances; cardboard materials, particularly food packaging; food materials including chicken heads and entrails; and other materials (Squires, 1983).

The cause of this event was traced to a 'highly unusual juxtaposition of happenings' including an oil spill, the explosion of two sewage sludge storage tanks, a number of pier fires, high river flow resulting from heavy rainfall and unusually persistent southerly winds. It was concluded that, although temporarily unpleasant, this event represented neither an immediate nor long-term health hazard. However financial losses related to the closures of beaches cost the area's economy $25–30 million (Squires, 1983). Although the public may have overreacted to this event, such happenings should warn us of the precariousness of our environmental situation. An event such as occurred in the summer of 1976 might have been considered unusual and of little threat. However, we continue to create situations in which such 'events' are practically inevitable. An increase in frequency or duration is all that is required to turn such isolated events into sustained environmental degradation.

In the summer of 1988, there were one or two reports of vials of human blood washing up on the beaches of Long Island, just east of New York City. These vials turned out to test positive for HIV. The public outcry was overwhelming, and was no doubt fuelled by the quite impressive media coverage. In the end, only a few vials were actually found, through the efforts of curious beachcombers and health officials alike. However, once again Long Island tourism suffered a blow for the summer. While the actual risk of exposure to HIV was quite small, the public outrage helped to focus attention on a larger, continuing problem related to the dumping of hospital waste (alleged to be going on in illegal sites), which was accumulating in significant amounts on beaches surrounding the New York Bight. These events happened to coincide with substantial fish kills, which were believed to be related to oxygen depletion in parts of Long Island Sound and other areas. Seemingly overnight large amounts of public concern (and research money) were poured into studies addressing pollution in New York's coastal waters. One of the problems with this scenario is that the obvious signs of serious pollution in the New York Bight and inner Long Island Sound had been ignored by the public for years, despite the warnings of scientists. (We have a photograph, taken in 1982, of a beach on Staten Island, NY that is literally covered in hospital waste, including such items as a completely intact intravenous unit). But apparently the public's perception of a serious direct threat to human health, supplied by a few vials of human blood, was required to get the public's attention. AIDS is without doubt a very serious disease, although the risk of contracting the disease by swimming on public beaches was, in this case, minuscule. Although the AIDS scare focused temporary attention on the serious problem of illegal dumping in the area's coastal waters, local residents report that the inappropriate disposal of hospital waste continues to present a serious problem in the New York City area.

While many ecotoxicologists appreciate the need to increase public awareness of environmental problems, we are not always as successful in this as we might like. Knowledge dissemination is often aggravated by a scientifically illiterate populace, for which the scientists themselves may be partly to blame. For example, several polls, including one circulated by the American Chemical Society, have indicated that although 70% of adult Americans want curbs on scientific activities, only 5% claim to understand basic scientific concepts or issues of science policy and only half of American 17-year-olds believe that science is useful (Callis, 1990)!

2.1.1 Role of the mass media

Studies in both the USA and Canada have strongly suggested that the mass media have played a major role in influencing public concern over environmental issues. Parlour and Schatzow (1978) outlined the debate over the media's role in influencing social attitudes, particularly in relation to environmental issues. These authors analysed the mass media coverage of various environmental issues in Canada between 1960 and 1972. Patterns of media coverage and public responses to surveys showed a tight coupling, suggesting that the media acted as major catalysts in generating and sustaining public concern. Unfortunately, the evidence also indicated that the media were very ineffective in educating the public about environmental problems.

The authors found that different issues, such as eutrophication and sewage pollution, showed striking differences with respect to the timing, scope and intensity of coverage. Summarizing trends in newspaper coverage over a 12-year period, they concluded that during 1960–5 major emphasis was on isolated local issues. During 1965–70 there was a shift to issues of national and then international concern. Analysis of the coverage of issues revealed a preponderance of sensationalized and exaggerated information. Furthermore, the scientific and technical content of the coverage was poor, 'with little attempt being made to increase the readers' understanding of these dimensions of environmental problems.' A lack of managerial initiative over environmental issues was blamed for the absence in the media of staff with the necessary scientific expertise for in-depth, objective and original coverage.

Parlour and Schatzow (1978) concluded that considerable subjective judgment by individual writers, reporters and producers outweighed the role of

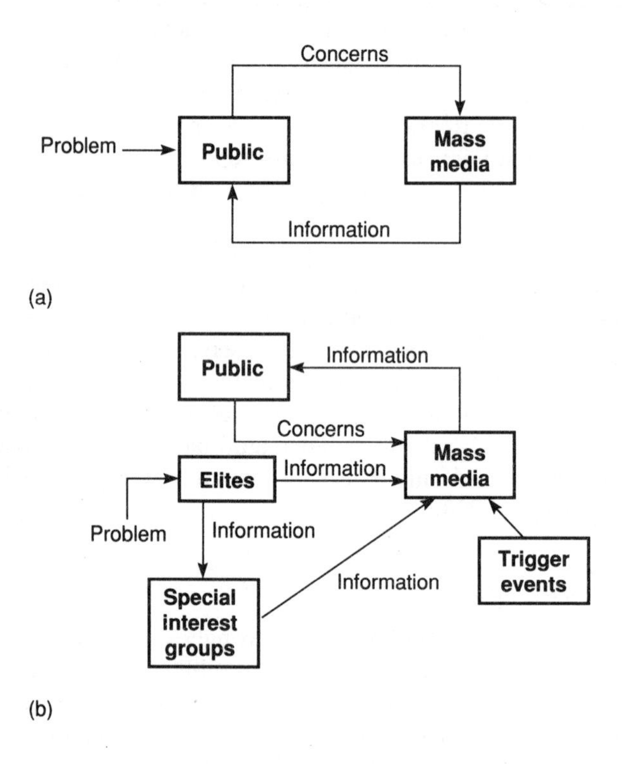

Figure 2.1 Models of the roles of the public and mass media in shaping environmental issues. (a) The media views itself as mainly responding to public concern about a perceived problem. (b) A more complex model in which concerns of special interest groups, intellectual elites and certain trigger events cause a response from the media. Reporting by the media serves to generate public concern, to which the media respond by increasing the level of coverage. Redrawn from Parlour and Schatzow (1978).

management in decisions regarding the importance of particular issues and of what would be in the public interest. Interestingly, the media were reported to believe that they were reacting to, rather than creating, public opinion. They viewed their role (shown in Figure 2.1a) as registering public concern and providing informational feedback. However, Parlour and Schatzow (1978) concluded that there was no evidence that the public was concerned about environmental issues before the media coverage. These authors suggested a more complete model of the media–public relationship (Figure 2.1b). This model emphasizes the importance of opinion leaders and the intellectual elites as playing a critical role in forming public opinions. In addition, certain trigger events (such as those described above in Long Island Sound) or the appearance of special publications or reports help to capture the media's attention, which is then communicated to the public, thereby increasing public concern. These authors suggested that as attention in the media shifts from environmental problems to other issues, levels of public concern decrease, irrespective of whether or not the problems get solved.

2.1.2 Public concern versus public behaviour

Both Canadian and American studies have marked the superficial nature of public concern over environmental issues which, interestingly, has not been accompanied by any significant changes in public behaviour. In Canada, which is perhaps second only to the Scandinavian countries in its level of environmental awareness, concern for the environment appears to have developed as a leisure issue for the middle class (Parlour and Schatzow, 1978). The evidence suggests that education is the prime determinant distinguishing concerned from unconcerned groups. However, even among concerned groups, the authors suggested that the lack of associated behavioural changes may be related to the perception by the public of their inability to improve environmental quality through individual actions. For this reason, voluntary measures to improve environmental conditions, such as voluntary recycling or waste separation, are more likely to be successful in smaller countries or at the local community level than at a large-scale national level.

The importance of public concern over environmental problems extends to the way in which public funds are spent. The allocation of public attention, and funds, in response to the latest perceived crisis, is not only inefficient, but prevents the development of coherent and long-term research strategies. When the crisis disappears, so does the funding. One of the things that frustrated scientists in trying to assess the significance of hypoxic events in Long Island Sound in the late 1980s was that there were not enough data from previous years with which to compare their recent findings. There was some evidence that hypoxic events might have been a relatively common occurrence in the Sound, but no one could tell how common (or if and by how much they had increased) because, until the crisis ensued, there was little money available to make the necessary measurements. There must be a better way to deal with our environmental problems than this knee-jerk, crisis–response reflex. Are we as scientists failing in our task to educate the public, or is society failing itself? What can we do about it?

In the public eye, industry is often viewed as the 'bad guy' and primary source of society's environmental problems. For example, in the USA, regulatory efforts designed to cut air pollution have focused on industrial emissions, and in some cities where air quality remains poor, Congress is imposing prohibitions on industrial construction. The reason that such restrictions are not likely to solve the air pollution problem is that the bulk of emissions are released by non-industrial sources. Statistics show that in a typical mid-western city, vehicles account for 33% of emissions; dry cleaning and printing for 23%; and solvents, paint and other surface coatings for 25%. Only 18% has been attributed to industry, and 'you can't reduce the size of the pie very much by reducing the size of the smallest piece' (Callis, 1990).

In a 1989 survey of beach litter in Halifax Harbour, Nova Scotia, it was reported that 62% was derived from domestic sources. The authors concluded that 'the litter problem in Halifax Harbour is created by the citizens of the area rather than by industry, the military or other sources.' The authors' recommendations for dealing with the litter problem included an education programme aimed at the general public as an important and effective step for controlling pollution (Ross *et al.*, 1991).

It appears that efforts directed toward finding solutions to environmental problems are more often based on political agendas than on effectiveness. As Callis (1990) wrote, 'The bottom line is that the public still does not understand that it is the activities of individuals that create the problems we encounter.' Parlour and Schatzow (1978) suggested that for many people the environment is an intellectual issue that has little or no bearing on their day-to-day existence.

An example from personal experience highlights the effective disassociation between humans and environment achieved in our modern societies. The State University at Stony Brook, where we did our graduate work, had many continuing education courses for adults. In order to fulfill the teaching requirements for our degree, one of us served as a teaching assistant for a course about Long Island Sound. The students in the course were a group of generally enthusiastic and highly motivated adults. In addition to attending lectures about the biology, physics, chemistry and geology of Long Island Sound, the students participated in a Saturday field trip to the local salt marsh where they were instructed on the marsh flora and fauna by their enthusiastic graduate teaching assistant. The instruction emphasized the historical and ecological importance of marsh ecosystems, the substantial destruction this habitat has experienced along the eastern coast of the USA, and the need to protect what little marsh was still left. The participants were quite interested to learn about marsh ecology and seemed to enjoy the trip. All of the students had spent significant portions of their lives on Long Island (most grew up there), all were obviously interested in the environment, but a number of them had never actually visited a salt marsh. How could this be? At its widest Long Island is only about 30 km with a north coast dotted with salt marshes. Surely one would stumble across a marsh at some point. Is it possible to spend one's life in an area and still be that out of touch with one's surroundings? Is it any wonder that vast areas of natural habitat are destroyed with little public reaction? Perhaps we should get more people to the marshes.

2.2. Deciding what to protect

Today only 3.2% of the Earth's landmass is legally protected, and if we consider that many areas are protected on paper only, the figure is even less (Erwin, 1991). In the past, conservation efforts have focused on a select number of species deemed to be in particular danger of extinction. The USA Fish and Wildlife Service has officially recognized 600 species as being threatened or endangered, but over 4000 species have been identified as candidates for endangered status (Soulé, 1991). In addition to protecting endangered species (or populations), the bulk of environmental protection has been directed toward efforts to minimize human contact or consumption of dangerous levels of pollutants and to prevent overfishing. Recently, protection priorities appear to be shifting from the level of the species to the level of the ecosystem, and the conservation of biodiversity is becoming of increasing concern.

Soulé (1991) identified five levels within a nested hierarchy which, in practice, guide environmental protection efforts. These are:

- ecosystems;
- assemblages (associations and communities);
- species;
- populations;
- genes.

Ideally strategies designed to protect ecosystems provide concomitant protection for all the lower levels. Soulé (1991) maintained that the debate over appropriate conservation approaches has oversimplified the differences among different tactics by exaggerating the distinction between species-level and ecosystem-level approaches. However, there are clearly situations in which strategies to protect single endangered species sharply contrast with those needed to protect the intact ecosystem of which they are a part (e.g., Alper, 1992).

Cairns and Mount (1990) suggested that functional attributes of very complex ecosystems, such as detritus mineralization, may be better indicators of stress on the system than are measures of the function of individual components (for example, species). In contrast, Schindler (1987) concluded that ecosystem functions would provide poor indicators of early ecosystem damage because feedback mechanisms act to conserve function. Both the development of more relevant research programmes and the implementation of more effective management strategies will require greater understanding of how structural and functional properties of individuals and populations are coupled (or uncoupled) to ecosystem structure and function (Chapter 6). Moriarty (1983) claimed that, 'In essence, our concern is with the size, structure and distribution of populations of individual species, and ideally we would monitor the impact, if any, of a pollutant on all species within a community.' If the goal is, in fact, to protect particular species, then our criteria are somewhat simplified, at least in theory. We may design our standards so that populations of protected species neither increase nor decrease as a result of human activities. Attempts at fisheries management have shown that, in practice, it can be quite difficult to protect even single species or populations, but the point is that our goal is quite clear. Once we reach the level of ecosystems our goals are much less obvious. Which functional attributes of ecosystems are the most indicative of a stressed system? Are these same attributes also the most important in other respects? For example, we may decide that the rate of detritus mineralization is an attribute that is worthy of protection. Such a decision should be based on fairly detailed knowledge of the detrimental effects resulting from an increase or decrease in the rate of mineralization (or respiration or photosynthesis, or any other variable we choose). The available evidence suggests that pollutants tend to produce similar changes in many different ecosystems, in particular, they tend to simplify the structure of both plant and animal communities (Moriarty, 1983; Rapport *et al.*, 1985; Gray, 1989). The consequences of such changes (in terms of continued ecosystem function, the production of goods and services etc.) need to be more fully evaluated as part of the development of coherent management and protection strategies.

It is apparently difficult, as Costanza *et al.* (1991) noted, to remember that nature is the economy's life support system. One of the problems we face, as scientists, is trying to persuade society to balance largely immeasurable attributes against hard economic facts or predictions. Lewis (1980) argued that biological considerations only receive priority when there is no strong economic conflict or when they coincide with other coincidentally beneficial requirements.

In an interesting and articulate analysis of our chemical dilemma, Milbrath (1991) designated 'life in a viable ecosystem' as the core value of society. He wrote,

> Some people seem to believe that according equal respect to all life somehow diminishes the importance of their own life. Such a conclusion is incorrect; rather, the point is that our own lives are inseparable from all life in the earth community; every time we diminish other life we diminish our own.

The economic value of ecosystems is often phrased in terms of services provided. 'Ecosystems can be thought of as producing goods and services for society. Goods are produced in the form of fiber and food, while services are produced in the form of assimilating waste materials, and providing recreation and aesthetics. Any impairment of this production might be considered unacceptable.' (Neuhold, 1986). Certainly, economic arguments for the protection of species and ecosystems can and have been made. From an economic perspective, species protection can be given a value based on:

1. knowledge;
2. potential for future sources of natural products;
3. physiological or biochemical uniqueness;

4. cultural reasons;
5. importance in maintaining the global climate and geochemical distribution of essential materials.

As Soulé (1991) noted, conservation strategies that protect the utilitarian value of ecosystems (such as water purification and storage, the promise of life-extending pharmaceuticals etc.) reflect the anthropocentric orientation of most societies. He contended that such narrow utilitarian arguments are politically weak because they focus on long-term benefits to society as a whole rather than immediate promises of survival and economic gain. Giddings (1986) suggested that most people would agree on the need for clean food and water and the need for restrictions to protect economically and recreationally important species. However, he argued, that many people 'would object to imposing regulations to preserve structural and functional properties of ecosystems, or to protecting plants and animals that are familiar only to botanists and zoologists ... It would be difficult, in our society, to argue convincingly in favor of protecting ecosystem properties for their own sake.'

Nevertheless, much current legislation is designed to protect the integrity of ecosystems (Emery and Mattson, 1986). Ecotoxicologists, in all branches of study, are devoting great efforts to overcoming the problems in accurately measuring the effects of pollutants on ecosystems and in extrapolating from laboratory test to ecosystem response. If, as Giddings (1986) suggested, societies do not value ecosystems for their own sake, then we have two choices. We can eliminate ecosystems from the focus of environmental protection and concentrate all of our efforts (with probably a much higher success rate) on understanding and protecting a few selected populations of lobsters and rainbow trout. Assuming that such a solution is unacceptable from a scientific as well as a societal point of view, the only long-term alternative is to develop convincing arguments for the value of ecosystems that are backed up by solid scientific data.

Wolff and Zijlstra (1980) suggested ranking the functions provided by given ecosystems, such as fisheries, research, recreation etc., with regard to their importance, to facilitate regulation of ecosystem use such that low-ranked functions are prevented from impairing higher ranked activities. Kinne (1980) argued that protection of ecosystems or parts of ecosystems for 'management of use-interests is insufficient and dangerously short-sighted from an ecological point of view.' He contended that environmental safety limits 'must be a matter for scientists free of direct commercial interests and pressures.' As a case in point, he cited our experience in fisheries management. Sustainment of fishery harvests are inadequate criteria for assessing ecological pollutant effects. As fishermen obtain their living from the sea, they are perhaps unavoidably more interested in short-term profits than in long-term environmental protection. As the requirements for successful fisheries conflict with other uses of the marine environment such as industry, recreation etc., decisions based on 'use-interests' will tend to favour the loudest voice. This may or may not be the wisest voice.

2.2.1 Protecting biodiversity

One of the major problems with current attempts at environmental protection and conservation is the lack of a coherent scientific framework. Concerning the preservation of biodiversity, Erwin (1991) observed

> There is no unified scientific method behind conservation strategy that addresses the nature and quantity of biodiversity, nor what it means environmentally either to save it or lose it outside direct human interest. In fact, there is little altruism or science in the fight to save the rain forest, the spotted owl, or the Antioch blue butterfly. Rather, politics and economics weigh heavily on most decisions. It seems that degradation and conversion of the environment is proceeding so rapidly that getting something preserved – anything at all – is acceptable regardless of the yardstick. Worst of all is that legitimate arguments within the scientific and conservation communities allow decision-makers an out in politically difficult choices. In order to supplement positive conservation practices and provide an alternative to negative ones, an effort to establish a sound scientific underpinning must be made. Scientific rationale may transcend cultural changes through time, whereas economic and political grounds certainly will not.

A fundamental reason for protecting ecosystems for their own sake is that we just do not understand enough about how and why they function, and thus cannot predict how damaging them will affect us. Probably the most well-known argument for protection of entire ecosystems involves the tropical rain forests. The rate of loss of rain forests, which have been reduced to about 55% of their original cover, approximately doubled between 1979 and 1989 (Ehrlich and Wilson, 1991). It has been estimated that deforestation will probably eliminate almost all of the tropical forests outside of protected areas by 2100 (Houghton, 1990). The destruction has been occurring at such a rate that species are driven to extinction before they have even been discovered. A conservative estimate suggests a loss of 4000 species each year due to tropical deforestation, but the actual figure may be two to three orders of magnitude larger (Ehrlich and Wilson, 1991). We have only the vaguest idea of what it is we are destroying so effectively.

The development of agriculture has reduced the number of plant species on which we depend to those most suitable for domestication. Approximately 78% of the world's total food production is derived from 15 species of plant (Barrett, 1981). Not only do we depend on the productivity of these crops for our survival, but many cultivated varieties are so specialized that they are unable to survive without our help. The desire of breeders to achieve uniformity in their plant crops and animals has had a huge influence on our agricultural practices (Barrett, 1981). Initially, uniformity seems to have been desirable for philosophical reasons. However, as farming methods became more sophisticated, it became increasingly important that crops would respond predictably (uniformly) to the practices used. In part, ensuring a uniform monoculture became of great practical importance. However, one of the consequences of achieving genetic uniformity, for example, among plants in a field, is that the susceptibility of plants to disease or parasitic infections is increased. As agricultural practices also usually involve increasing the density of plants, the potential for epidemics is magnified. The dangers of monoculture have begun to be appreciated, and some more recent practices have involved a shift from single varieties to mixtures. However, the ideal of achieving uniformity is in some cases so ingrained that attempts to change current practices have met substantial resistance. Barrett (1981) provided an interesting example for the case of cotton.

> In order to promote the foundation of a high quality cotton industry in the San Joachim Valley, California, the state legislature passed a statute in 1925, requiring that just one variety, Acala, be grown. This statute made it an offence [*sic*] to even possess another variety with the intention of planting it. In the first year of the operation of this law, the area of cotton doubled and has now reached 5×10^5 ha. However by 1970 15–20% of the area was severely infected with *Verticillium* to which the variety Acala is very susceptible. In the infected area farmers averaged only 2.7 bales/ha compared with yields of the order of 7.7 bales/ha in uninfected areas. But they are still required by law to grow this susceptible variety. Many unsuccessful attempts have been made to alter the law: the poor yields have been blamed on "poor farming practices" and "worn out soils"; this may be true but the existence of the law precludes any flexible response to such problems by the growers.

Although it may not always be true that the most diverse ecosystems are the most stable, we know from our experiences in agriculture that monocultures are highly unstable and are more prone to decimation by pests, disease, or environmental changes. At worst, an 'ecosystem' which contained only those species of immediate human economic or recreational value would be highly unstable, and thus difficult to manage. It would, no doubt, have countless unforeseen consequences, with regard to human health and happiness, and it would have a very limited potential for future exploitation. At best, the intentional creation of a world consisting of wheat, some cows, a few trout or maybe striped bass, the incidental rat and cockroach (which we probably could not drive to extinction if we tried) and of course our best friend, the dog, would cast serious doubt on the supposed superior intelligence and imagination of *Homo sapiens*.

As discussed above, conservation efforts have largely been directed toward

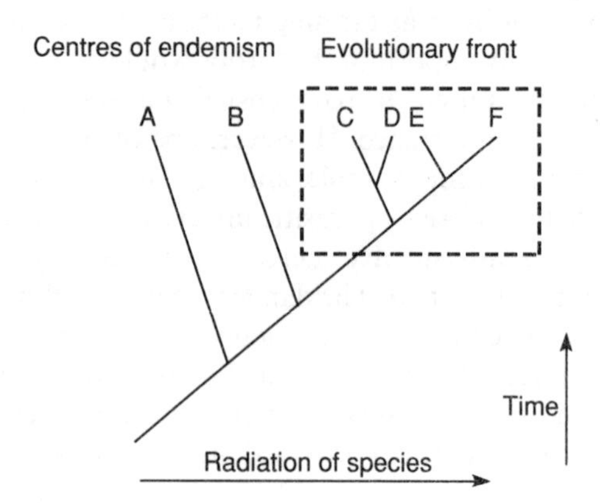

Figure 2.2 Taxonomic relationship among six species having a common ancestor. Although current conservation efforts place highest priority on protecting endemic species such as A and B, species C, D, E and F hold the most promise for continued evolution of this line of biodiversity under natural conditions. Redrawn from Erwin (1991).

target species that are of economic, recreational, aesthetic, or cultural value. Such conservation strategies may be ineffective in the long term because they are not specifically designed to protect evolutionary processes and environmental systems that will generate future biodiversity (Erwin, 1991). The analysis of phylogenetic relationships can be used to identify evolutionary fronts (centres of radiation) the protection of which Erwin (1991) argued is essential for maintaining future biodiversity. For example, current conservation strategies would place the highest priorities for protection on species A and B (shown in the cladogram in Figure 2.2) because these species may be endemic, rare and geographically restricted. Erwin (1991) posited that species A and B, although worth protecting for the evolutionary information they carry, have demonstrated no radiation and are predictably on their way to extinction. In contrast, species C, D, E and F are more widespread, and may even be considered weedy, but are most likely to experience radiation and hold greater promise in terms of future evolutionary change.

Following the recently articulated view of Woodwell (1991), we suggest that a primary area of concern should be the role of pollutants in causing biotic impoverishment or **the reduction in the capacity of ecosystems to support life**. By defining damage to an ecosystem as a decrease in viability or life-supporting capacity we can make informed decisions about undesirable pollutant effects. Although this may provide us with a reasonably clear goal, it is obviously an extremely tall order. How can we go about fulfilling it? One approach that we feel would be extremely useful would be to focus on pollutant-induced effects that influence diversity at different levels of biological organization. Specifically, properties of primary interest would be genetic diversity within populations, diversity (the number and kind) of species in a community, and diversity of available habitats over space and time. A defensible working hypothesis is that maintaining diversity both within and among the many levels of organization will tend to maintain system viability and ensure the provision of key services to life. Intraspecific genetic diversity keeps evolutionary options open and is thus inherently worthy of study and conservation. Second, an understanding of pollutant effects on interspecific diversity is also important. This is because complex interactions among many species are often key determinants of ecosystem function. Thirdly, the diversity in abiotic factors that control habitat structure over space and time can limit the kind and abundance of organisms that can survive in an area. The number and variety of niches available in heterogeneous habitats favour higher levels of biotic diversity.

It is generally appreciated that highly diverse ecosystems, such as the tropical rainforests, have yielded substantial immediate benefits to society, including various foods, medicines and industrial products. As our technical abilities to identify, isolate and synthetically mimic natural substances improves, the potential for even greater economic and health benefits continues to grow. Also of great importance, and perhaps the least understood, is the role of biodiversity in maintaining the essential functioning of ecosystems. Regulation of atmospheric gases, recycling of rainfall and nutrients, pollination, protection from pests, maintenance of soils and decomposition are all important functions of ecosystems which rely on complex interactions among their component parts. Ehrlich and Wilson (1991) aptly noted that virtually all human attempts to substitute large-scale inorganic alternatives for natural ecosystem functions, for example, substitution of synthetic pesticides for natural pest control, inorganic fertilizer for natural soil maintenance, chlorination for natural water purification, dams for flood and drought control, have proved inferior.

Although the destruction of tropical rain forests and the expansion of agricultural monocultures are not ecotoxicological problems, *per se*, their consequences, namely the destruction of habitat and the shift to one or a few dominant species, are analogous to pollutant-caused effects. Whereas ecotoxicology specifically addresses changes in ecosystem structure and function resulting from chemical pollution, its primary objectives, to identify, quantify and ultimately control various human impacts, is shared with other branches of environmental study. It has been emphasized that only by developing management approaches that integrate pollution control with other sources of anthropogenic effects (for example, deforestation, species extinction, ozone depletion etc.) can we stand even a chance at solving the complex environmental, social and political problems that we now face (Cairns and Pratt, 1990). We will proceed to address some of the major difficulties we encounter in meeting these objectives.

2.3 Deciding what is significant

Before we can decide whether effects caused by pollutants are adverse or unacceptable, we have to be able to determine with some degree of certainty that an effect has indeed occurred. The application of statistical techniques can provide powerful and objective tools for data analysis. Like any tool, the usefulness of statistics relies on its proper application. Statistical tests are applied with the intention of determining whether an effect is significant. But what is a significant effect? Those unfamiliar with statistical methods of data analysis sometimes view with suspicion results that are found to be statistically significant because of the fear that such results are not also biologically significant. Those familiar with statistical methods will recognize this situation as a type I error. For any given statistical test there is a stated probability, chosen by the investigator (but conventionally 1 in 20), that an effect will be observed when in fact none exists. An alternative outcome arises when an effect exists but is not detected by the statistical test, that is, the effect is biologically, but not statistically, significant. The probability of this kind of error, referred to as type II error, can also be controlled to some degree by the investigator. The importance of type II error has received less attention, but should be of particular concern to ecotoxicologists given the potentially serious consequences that can result when statistical tests fail to detect biologically significant effects. A more complete discussion of type I and type II errors follows in Chapter 4.

To this list we add two additional types of errors that are frequently encountered in studies related to environmental management and protection. We suggest that a type III error occurs when an effect is found to be scientifically, but not politically significant. Effects falling within this category, although deserving of study, may be neglected owing to the inability of scientists to obtain available research funding to study problems that are deemed to be 'unimportant'. For example, government regulatory agencies often have been known to find the study of substances that they have already regulated or banned, such as tributyl tin and DDT, to be undeserving of further study, even though biological effects of such substances may continue to be important for a long time after regulation has commenced. [According to a 1990 report published by Joint Group of Experts on the Scientific Aspects of Marine Pollution (GESAMP) on the state of the marine environment, levels of DDT are still surprisingly high in many areas (J. Gray, personal communication).] The alternative situation has also been known to occur; that is, effects or problems which are of minor scientific significance have for various reasons achieved great political significance. Such problems may receive a great deal of attention regardless of the fact that their solutions may contribute more to the success of a few careers than to advances in scientific understanding (type IV error). The unsuccessful allocation of funds and regulatory efforts towards reducing industrial emissions (discussed above), in order to reach air quality standards, may provide an example of this type of error. Despite the fact that individual automobile emissions had the most significant effect on total levels of air pollution (Callis, 1990), greater attention was focused on reducing the more politically important industrial emissions.

This list could be extended to include other types of significance, but by now the message should be clear. As ecotoxicologists, our primary concern is with the scientific importance of pollutant effects. However, in ecotoxicological studies, more than in many other branches of science, the direction and interpretation of research questions are influenced by non-scientific issues. It is necessary to remain cognizant of these issues, not the least because they can lead to an unrecognized loss of objectivity and to ineffective research programmes.

2.4 Deciding what is acceptable

Ecotoxicologists aim to assess and predict the effects of pollutants on ecosystems, that is, we want to be able to decide if a given ecosystem is being or has been altered as a result of pollutant exposure, to determine how serious the resulting effects are and to forecast the likely fate and effects of using or releasing a specified amount of a substance in the future. Not only has there been substantial doubt as to the best way to achieve these aims, but some are beginning to question whether or not these goals are fundamentally unreachable.

Current legislation in the USA and Europe is supposed to protect ecosystems against adverse changes in diversity, productivity and stability. Emery and Mattson (1986) contended that such legislation makes several assumptions.

1. Ecologists can define and identify stable ecosystems.
2. Stability and response to disturbance are predictable properties of ecosystems.
3. Interpretations of ecologists will be convincing in legal contests.

In attempting to define what pollutant-caused changes are unacceptable, Stephan (1986) proceeded through a series of possible definitions from the most protective (that any change is unacceptable and therefore input should be limited to zero), through a somewhat more practical definition (that all adverse effects are unacceptable). He noted that this leaves us with the problem of defining 'adverse' and that neither of these definitions takes into account the possible benefits (to humans) associated with the change. Thus, he argued for a definition of 'unacceptable' that is not based on purely scientific criteria.

> People are the ones who make, use, and dispose of chemicals and products, and they are the ones who must be involved in choosing among options concerning production, use, and disposal. Choosing among options involves decisions concerning acceptability based on value judgments. People have different attitudes concerning what is acceptable and unacceptable, but it is important to realize that these attitudes are based on various personal value judgments.

As stated by Stephan (1986), 'A major goal of applied aquatic toxicology is to be able to make a useful prediction concerning whether or not a specific addition of a toxic agent to a particular aquatic ecosystem will cause any unacceptable effect on that ecosystem.' To meet this goal implies that aquatic toxicologists are able to define 'unacceptable'. This suggests two important questions.

First, does there exist a set of reliable criteria that can be used to determine whether a given effect is 'unacceptable'? Ideally, we would want to have a more precise measure than just acceptable vs unacceptable, so we would have to have criteria that allowed us to characterize the degree of unacceptability of a given change (an unacceptability index). This has led to the widespread introduction of various quality criteria that attempt to provide legislatable measures for assessing the condition of water, soil etc.

Second, assuming ecotoxicologists can define what is unacceptable from a scientific perspective, consideration of non-scientific issues can greatly influence what is deemed unacceptable. Whether human society has the desire, stamina and enlightenment to tackle these tough issues effectively is unknown.

2.5 Obstacles to scientific understanding

Many factors complicate the detection of pollutant-caused effects. In making a scientific assessment of the effects of pollutants on ecosystems, the ecotoxicologist is faced with several difficulties (Kinne, 1980). He or she must be able to:

1. differentiate between natural and pollution-caused changes;
2. determine the effects due to pollutants acting in concert;
3. exactly identify and correlate cause and effect;
4. select reliable and readily measurable criteria which facilitate a precise characterization of ecosystem state.

Perhaps as a result of these difficulties, approaches designed to control pollution have focused on limiting effluent inputs even though the ultimate criteria for environmental evaluation are the responses of populations and communities in nature (Lewis, 1980).

The null hypothesis of pollutant studies can be stated as follows: 'that there is no pattern of difference among samples that is correlated with the known amounts of pollutants' (Underwood and Peterson, 1988). A number of biotic and abiotic factors hinder our ability to adequately test this hypothesis.

It has been recognized that different species or trophic levels exhibit different spatial and temporal periodicities, which are controlled by some combination of biotic and abiotic properties of the system (Maurer, 1987). Thus, the suggestion that single species dynamics can be used to extrapolate to community or ecosystem response has met with considerable doubt. Giddings (1986) argued that 'the threshold of persistent, significant damage to an ecosystem is likely to be rather sharp. Determining the safe level of exposure should not require subjective decisions about which effects are ecologically significant and which are not: above the safe level, effects are likely to occur throughout the ecosystem.' However, recent data on the effects of oil pollution on marine benthic fauna suggest that, indeed, rather subtle effects can be detected using objectively defined criteria (by multivariate analysis of species composition changes) and also that effect thresholds are diffuse, rather than sharp, but nevertheless very clear (Gray *et al.*, 1990).

Natural populations of organisms fluctuate in both time and space. This feature greatly limits our ability to detect changes in abundance or distribution that result from pollutant impact. Emery and Mattson (1986) defined ecosystem stability as 'the maintenance of structural and behavioral order as measured by steady states, continuous cycles, rhythmic fluctuations, or stable oscillations of appropriate ecosystem characteristics ... Disturbance of such order would be observed when any of these characteristics display evidence of chaos, viewed as a departure from natural or desired trajectories.' Williamson (1987) reminded us that 'no community has been stable to evolutionary change', and thus community stability can only be assessed with reference to stated time scales.

Much emphasis has been placed on the study of species diversity and abundance. In Chapter 3, we discuss some of the problems with diversity indices. Primarily, they have been criticized as being too insensitive to pollutant perturbations and have no theoretical distribution against which deviations can be objectively tested (Gray, 1980). Inherent in these measures is the assumption that the more diverse a community, the greater the number of compensatory interactions, and thus the more stable it should be to perturbations. However, ecological theory casts doubt on this assumption, and both May (1973) and Roughgarden (1977) have presented separate arguments for an inverse relationship between diversity and stability. Another complicating factor is that some, but not all, natural populations of organisms can have great capacities to withstand a substantial reduction in numbers. To what extent populations of a single species differ in this regard has, to our knowledge, not been quantified.

Moriarty (1983) argued that our ability to predict the ecological effects of a pollutant is limited because often more than one compound is involved, the pollutant's distribution in the environment is not uniform, relationships among exposure, effects and amounts of pollutant in organisms are complex and vary within and among species, and the role of interactions among individuals and species is poorly understood. The effects of stress on communities have included structural but no functional change (e.g., Rapport *et al.*, 1985); functional but no structural change (e.g., Davies and Gamble, 1979); and both structural and functional change (e.g., Wängberg *et al.*, 1991). Klerks (1987) and Klerks and Levinton (1989) noted that the evolution of resistance in certain species exposed to pollutants can complicate our efforts to predict the fate and effects of pollutants in food webs. Increased resistance has been associated both with decreased accumulation of pollutant, as well as with increased accumulation by resistant organisms. In such situations, predictions of pollutant behaviour in food webs by laboratory bioassays will not be valid.

Underwood (1989) identified three key attributes of populations that will influence the population's response to perturbations, namely inertia, stability and resilience. The inertia of a population will determine whether or not it will respond to a given size, type or frequency of perturbation. The greater the inertia, the greater the stress needed to elicit a response. Thus, this parameter can be used to determine whether or not a given perturbation acts as a stress. The stability of a population is defined as a measure of its rate of recovery following the appearance of a given level of stress. Resilience refers to the level of stress (for example, measured as the number or percentage reduction in population size) from which a population can recover. Populations that can reattain equilibrium following a large perturbation are said to be resilient.

These attributes provide a useful conceptual framework for generating hypotheses about how natural populations might respond to stress from pollutants. However, implicit in these definitions is the need to be able to measure equilibrium states and to distinguish these from nonequilibrium states. But as Underwood (1989) wrote,

> There are great conceptual difficulties in definitions of equilibria in natural populations. The organisms and their food, predators and pathogens, fluctuate widely in time, and differ from place to place because of a host of ecological variables. Whilst these are usually responses to perturbations of some sort, they create a background against which the detection of equilibria becomes difficult, if not impossible.

How do we overcome such difficulties? Can we measure perturbations for nonequilibrium populations? In theory, it should be possible to measure a population, over time, before and after a perturbation, and to compare changes between sampling times without assuming that the population was in equilibrium. An alternative approach would be to compare populations from perturbed and unperturbed sites. However, the validity of both of these approaches is very sensitive to the sampling design. Underwood (1989) outlined potential statistical and logical pitfalls associated with such approaches.

Giddings (1986) hypothesized that 'safe exposure levels for ecosystems are usually near the lower end of the chronic toxicity range for single species.' Contrary to this idea, Davies and Gamble (1979) suggested that intact ecosystems are actually more robust to perturbations than are isolated individuals taken from it. Although Underwood (1989), Underwood and Peterson (1988) and others emphasized the difficulties in determining exactly at what point a community or ecosystem is significantly impacted by pollutants, Giddings (1986) claimed that determining safe levels (for contaminants) should be relatively easy because the safe level 'represents a distinct discontinuity in the gradient of ecosystem response to increasing exposure.' Stern and Walker (1978) maintained that in designing biological testing for decision-making purposes, 'The most defensible approach is the use of a combination of tests involving the most sensitive ecological indicator organism and life stage commonly available.' To what extent certain species (or populations or communities for that matter) are consistently sensitive or consistently tolerant to a wide range of pollutant perturbations is open to debate. Another relevant question with regard to sensitive species is, 'Why are they sensitive'? Can we identify some general attributes that are common to those species we deem to be generally 'sensitive'? Kinne (1980) postulated that species tolerant to variations in abiotic factors tended to show low tolerance to biotic factors such as competition and parasitism, and vice versa. He termed the former species 'euryplastic' and suggested that such species generally occupy habitats characterized by physical instability, for example estuaries, and tend to have lower rates of growth, metabolism and reproduction. Kinne (1980) argued that species and ecosystems with low rates of metabolism and turnover are less vulnerable to pollution than are high metabolic-rate systems. This argument might suggest that estuarine organisms will be more tolerant of abiotic stress in the form of pollutants. However, several pages later, Kinne (1980) went on to say,

> Ecologically, estuaries and coastal waters tend to support ecosystems under considerable abiotic stress (e.g., extreme intensities and variations in temperature, salinity, oxygen level, turbidity). Hence, additional stress due to pollution may quickly lead to critical impact levels. For these regions, safety levels must be set high – much higher than laboratory resistance tests with euryplastic estuary inhabitants would generally indicate.

Estuarine organisms are no doubt in substantial danger from pollution, largely because of the high loads that such systems receive. However, determining whether organisms from such abiotically stressful habitats are, as a general rule, more tolerant of chemical stress than their counterparts from abiotically stable habitats will require that the tolerances of a variety of species, chemicals and habitat types be examined under field conditions. Studies by Fisher *et al.* (1973) and Fisher (1977) found that phytoplankton clones isolated from the Sargasso Sea were less tolerant to polychlorobiphenyls (PCBs) and several novel organic compounds than were clones isolated from temperate zone estuaries. Fisher (1977) suggested that organisms which have evolved in and adapted to physically variable environments would, because of their adaptations, be better able to tolerate exposure to toxic compounds than would morphologically similar organisms adapted to stable environments.

In ecology, as in ecotoxicology, the role of variability (genetic and phenotypic) among individuals in a population has been largely ignored (Kareiva, 1989). Bennett (1987) articulately exposed 'the tyranny of the golden mean' that has characterized comparative physiology and physiological ecology and led both fields to overlook the valuable information that can be derived from interindividual comparisons. Although many populations are experiencing reductions in genetic diversity, the implications of such reductions on population dynamics or on population responses to stress are poorly understood. In ecotoxicology, as in physiology and ecology, the tendency has been to view variability among individuals as noise and as something obscuring the 'true' response. Thus, great efforts have been devoted towards standardizing test systems by limiting genetic and phenotypic variability among test organisms. At best, such an approach misses valuable information; at worst, it can lead to the establishment of inappropriate regulations. We suggest that preservation of such variability and a more direct analysis of it will provide new and valuable insights into ecotoxicological phenomena and will lead to more realistic assessments of the sensitivities of natural populations to chemical pollutants (Chapter 4).

2.6 Ecosystem health: a fit phrase?

Although both 'community' and 'ecosystem' are concepts fundamental to our understanding of ecological processes, these terms have been inconsistently used in the literature. Anderson and Kikkawa (1986) traced the historical development of the concept of community and the emergence of theories of community ecology. Of those definitions in the literature, we find Ricklefs' (1979) definition of a community as 'an association of interacting populations, usually defined by the nature of their interaction [and] of the place in which they live' and Whittaker's (1975) definition of ecosystems as 'functional systems formed by communities and their environments' to be particularly useful. In practice, communities are difficult to define. Sometimes they are relatively easy to identify as discrete units and sometimes they appear to overlap and blend one into the next. For this reason, many ecologists often resort to the term assemblage rather than community.

Loehle (1987) warned of the problems that can arise in science when basic terms, such as 'community' or 'stability', are ambiguously defined. For example, he noted that in early theories of ecological diversity, the term 'diversity' was used to mean total number of species, landscape diversity, number of species per number of individuals, as well as various theoretical indices.

Because there are no universally applicable criteria for operationally defining such concepts as 'ecosystem', Loehle (1987) argued that theories based on such concepts are bound to lead to inconclusive debates. Uncertainty in such definitions can lead to serious problems when decisions have to be made regarding ecosystem or community protection and management strategies. An excellent example of the use of ambiguously defined concepts is provided by the phrase **ecosystem health**.

The terms **well-being** or **health** very often appear in the environmental literature to describe the state of communities or ecosystems. Usually this usage is in reference to some property of communities or ecosystems that is considered to be in need of protection. The term 'well-being', as commonly used and defined in the dictionary, is 'the state of being happy, healthy, or prosperous'. In turn, 'health' is defined as 'the condition of being sound in body, mind, or spirit, especially: freedom from physical disease or pain'. According to this definition, it would clearly be appropriate to discuss the health of an individual organism. The degree to which an organism is unhealthy describes its potential for premature mortality or reduced reproductive success. An organism's health is likely to influence its contribution to the system (be it population, community or ecosystem) via its productivity and interactions with living and nonliving components of the system. Health presumably also has an influence on the organism's reproductive abilities, and is therefore associated with fitness (defined here as the organism's genetic contribution to the next generation (Maynard Smith, 1989)). On the other hand, if there are tradeoffs between the resources devoted to stress-tolerance mechanisms and the resources devoted to growth or reproduction, individual health (that is, the degree to which normal functioning is maintained under pollutant exposure) may be inversely related to fitness (Holloway *et al.*, 1990). In any case, the meaning of the term 'health', as commonly understood, is relatively unambiguous when applied to individual organisms. Objective measures of organism health exist. Such is not the case for ecosystems.

Pollutants can cause lethal and sublethal effects on individuals, but not on populations or communities (Moriarty, 1983). Populations of organisms can be neither healthy nor unhealthy, they can only increase or decrease in number. Whereas measuring a change in population number may provide an appropriate criterion for determining a pollutant-caused effect, it usually has an unknown, and often variable, relationship with the health of the individuals comprising the population. The term health cannot be used to describe the state of ecosystems in the same way that we use it to describe the state of individual organisms. However, rarely is health explicitly defined when discussed in an ecosystem context. There have been a few (e.g., Rapport *et al.*, 1985) who have explicitly made the analogy between individual health and ecosystem health. Rapport *et al.* (1985) proposed that detecting ecosystem unhealth could proceed, as in medical practice, by comparing certain ecosystem functional indicators for deviations from normal values. They suggested that changes in parameters such as nutrient cycling, primary productivity, species diversity, species composition and species distribution could be used as measures of ecosystem health.

There are several practical problems involved in this approach, not the least of which is determining what a normal value for a given ecosystem should be. There is no question that certain generally (though not universally) consistent changes occur in ecosystems influenced by human activities (e.g., Rapport *et al.*, 1985; Schindler, 1987; Gray, 1989). These changes may be undesirable because they are indicative of an imbalance between man and nature. Such a conclusion is very different, however, than concluding that a thriving ecosystem dominated by a few species of opportunists is in any way unhealthy. An ecosystem may be quite different following exposure to a pollutant, relative to how it appeared before exposure. It may cycle nutrients at a different rate, it may be more or less stable, it may be more or less diverse. Although we may very clearly be able to decide whether it is more or less desirable, we cannot conclude that it is more or less healthy.

In contrast to Rapport *et al.* (1985), we suggest that individual health can be more objectively defined (recognizing that defining health is one problem and measuring it is another) and can be directly related to the probability for continued survival and reproduction. The criteria for defining ecosystem health are not as clear and, for purposes of directing ecosystem management strategies at least, appear to be less important than criteria based on desirability. Although the idea of a healthy ecosystem appeals to our aesthetic sense, it is a misleading term because it presumes that all healthy ecosystems are desirable. A thriving ecosystem dominated by sulfur bacteria can function extremely well and its component members be quite healthy, although there are few who would argue for its preservation in environments formerly dominated by coral reefs. In short, the term ecosystem health is an inappropriate concept. As this term has been generally used, it has presupposed that objective measures of ecosystem health exist and that such criteria provide the logical basis for environmental protection efforts. Our major concern is that changes related to perturbations have been inappropriately associated with the concept of health, and that this association has not served to clarify our understanding of pollutant impact on ecosystems.

We suggest that the term **ecosystem damage** may be less ambiguous than health (or unhealth) and has the advantage of incorporating the role of value judgments. In standard dictionary usage, the word 'damage' refers to the process of spoiling the condition or quality of something and the harm or loss that results. Of course, such a definition leaves us with the problem of determining the condition or quality of an ecosystem and of deciding which pollutant-induced changes are acceptable and which are not. Solving this problem is made somewhat easier by the fact that we have essentially two options with regard to environmental protection and management strategies. Our efforts can either be directed toward making ecosystems 'the way they were before human activities impacted them' or 'the way we want them to be'. The first of these options may not be possible for most cases. Thus, left with the second option, it is clear that the role of value judgments needs to be more widely recognized and accepted as a valid basis for ecosystem protection and management. Our ability to manage ecosystems such that they are in a state that we desire will require fundamental knowledge of their structure and function.

Friedland (1977) discussed the theme of 'values and environmental modeling'. He distinguished between 'facts', those statements, assumptions, beliefs and so forth, that are amenable, actually or in principle, to scientific testing, and 'values', those statements, assumptions, beliefs and so forth, that by their nature cannot be tested. He argued that both facts and values are essential ingredients in any decision.

> For no amount of facts, however extensive, is sufficient to enable a model builder or any decision maker to decide which course of action to take. Somehow, the alternatives available must be ordered in terms of desirability... . Values alone are the instruments with which decision makers are able to reduce into a unidimensional index of merit or worth the multidimensional information describing each alternative and the possible or likely consequences of its adoption.

Deciding what attributes of ecosystems we wish to protect from pollution cannot be based on objective criteria alone.

2.7 Decisions in risk assessment

2.7.1 Objectives and scope

Environmental protection and management strategies rely on the accurate estimation of the hazard, impact or risk of chemicals to ecosystems. The terms hazard assessment, risk assessment and impact assessment have been used inconsistently in the literature. Barnthouse *et al.* (1986) defined the goals of **impact assessment** as 'the identification of specific environmental issues related to a proposed project and the consideration of those issues in the project planning process.' Thus, **risk assessment** can perhaps be considered as a specific type of impact assessment, one that focuses primarily on the toxicological or ecotoxicological consequences arising from the development, manufacture or release of chemical substances. In contrast, impact assessment can include a wide variety of changes in, for example, water flow, temperature etc., that may alter in response to a specified human activity. Cairns *et al.* (1978) defined **hazard assessment** as a process whose goal is to provide information from which a judgment can be made regarding the safety of a substance or activity. It is a process involving both scientific data and some kind of value judgment. First a hazard assessment permits the potential for harm to be evaluated, and second, a value judgment is required to determine whether or under what conditions the risk is acceptable (Maki and Bishop, 1985). Two critical aspects of hazard assessment are first, deciding which toxicity tests will reduce the uncertainty of risk most cost-effectively and second, determining at what point the uncertainty has been sufficiently reduced to allow an informed professional judgment to be made (Cairns and Mount, 1990). Hazard assessment is considered by some authors to be a formal step in an overall risk assessment (Kuiper-Goodman, 1989). Alternatively, Maki and Bishop (1985) explicitly used the terms 'risk' and 'hazard' interchangeably. Volmer *et al.* (1988) defined **risk assessment** as 'a method for estimating both the magnitude and probability of deleterious effects of anthropogenic substances to the environment.' Given that assigning realistic probabilities is often not achievable, there may be little difference between hazard assessment and risk assessment in practice.

Risk is not directly measured, but rather calculated from other events, and as such, it involves a number of assumptions and uncertainties. For example, the risk of developing cancer from lifetime exposure to low concentrations of a chemical is estimated from data on exposure to higher doses for shorter periods of time. As defined by Scala (1991), risk assessment is 'a process whereby relevant biological, dose–response, and exposure data are combined to produce a qualitative or quantitative estimate of adverse outcome from a defined activity or chemical agent. Scala (1991) defined **hazard identification** as the first step in risk assessment. It is the process by which a substance is determined to be potentially dangerous for human health or the environment. Hazard identification is based on factors related to the potential mobility of a chemical in the environment, the potential (biotic and abiotic) reactivity of a chemical and the potential effects of the chemical on living systems (Bro-Rasmussen *et al.*, 1984).

Dose–response assessment, the next step in risk assessment, attempts to determine how dangerous a substance is, that is, to relate environmental concentration and effects on living systems (this step is referred to as hazard assessment by some authors, for example, Kuiper-Goodman, 1989). **Risk extrapolation**, the third step in risk assessment, involves extrapolating the toxicological data from one species to another, using safety factors or other methods, to arrive at an estimate of safe exposure. The extrapolation procedure requires that effects of high doses are used to predict effects at low doses, outside of the experimental range, and that effects on test species are used to predict effects on other species potentially exposed to the chemical. Uncertainties include selecting the appropriate dose-scaling factor for organisms of different sizes. It is often not clear which scaling factor is the best, and large differences in calculated doses can occur when different factors are used. For example, if body weight instead of surface area is used for scaling chemical dose from mice to humans, estimated average daily dose for humans is reduced 14-fold (Scala, 1991).

Finally, **exposure assessment** is the part of risk assessment that includes estimates of the intensity, frequency, schedule, route and duration of exposure, as well as of the nature, size and make-up of the exposed population. Estimates of the amount of pollutant to be produced, knowledge of its intended uses, its tendency to disperse in the environment, its potential to persist and be converted, and its consequences for the biota are all essential components needed to determine environmental risk. The overall risk is characterized by combining the information on hazard identification, dose–response and exposure to provide an estimate of the incidence of adverse effects in a specified population. Once the scientific data have been combined into an assessment of risk, additional factors, including policy, social and economic factors, technical feasibility etc. are considered in the selection and implementation of control strategies for risk management. Lloyd (1980) identified three specific objectives of risk assessment which may be emphasized to varying degrees.

1. Protect species of economic importance or to ensure acceptability for human consumption.
2. Protect biota on biological grounds or to maintain human amenities.
3. Ensure that a minimum amount of biological change occurs.

Scala (1991) outlined the elements of risk assessment and risk management used in human toxicology, and these are analogous to those involved in the assessment and control of environmental risks and are shown in Figure 2.3.

Another aspect of hazard or risk assessment involves a calculation of the tradeoffs involved in, for example, the release of a potentially hazardous chemical or the implementation of a particular management strategy. In order to calculate these tradeoffs, a system of common currency is needed with which we can express the various costs and benefits involved. This is often one of the more difficult aspects of any management decision. One example of how this problem can be approached is found in Kemp *et al.* (1977). These authors estimated the economic and ecological costs and benefits of constructing

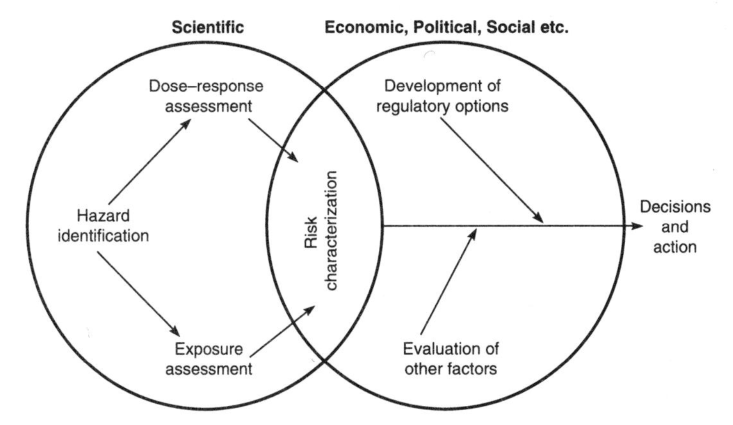

Figure 2.3 Basic elements of risk assessment and their relationship to each other. Adapted from Scala (1991).

cooling towers to reduce thermal discharge from a power plant by expressing all variables in common energy units.

2.7.2 Pollutant fate and effects

Ecotoxicological risk assessment explicitly attempts to relate the fate of chemical substances to their biological effects. Determining both the effects of chemicals (or processes) on the environment and the effects of the environment on chemical fate are critical to sound risk assessment and, in fact, provide a broad framework into which most ecotoxicological studies fall. We have noticed an unfortunate tendency among ecotoxicologists to treat fate and effects as separate entities, often to the extent that researchers focus exclusively on one or the other process with little exchange (and on occasion, downright animosity) between opposing camps. We stress (Chapter 5 has a full discussion of this point) that focus on the interdependency of pollutant fate and effects is both necessary and desirable. To divorce the two would be not only unwise, but impossible. For example, in mesocosm systems to which fuel oil was added as a pollutant, microbial degradation of hydrocarbons was found to increase after addition of the fuel oil (Lee *et al.*, 1977). This increase suggests a change in the microbial population towards an increase in the number of microorganisms able to use hydrocarbons as a carbon source. Thus, the **effect** of the fuel oil on the microbial community resulted in selection among genotypes or species which had a profound influence on the **fate** of hydrocarbons in the water column.

Fate is determined by a substance's intrinsic chemical and physical properties, by the chemical, physical and biological properties of the environment into which it is released and by the amounts and rates at which it enters an ecosystem. Among the intrinsic properties generally considered to be critical in evaluating a chemical's potential risk to the environment are its molecular structure, solubility in water and organic solvents, vapour pressure, particle size and absorption spectra (ultraviolet, visible). In addition, its tendency for sorption to solids, effects of dispersing agents if used and volatility can also influence chemical behaviour. In aquatic systems important (physicochemical) environmental parameters include surface area, depth, pH, degree of turbulence, percentage carbon in the sediments, temperature, salinity and suspended sediment concentration (Cairns *et al.*, 1978). A number of processes can lead to alterations in the toxicity of chemicals after they have reached the environment. These include dilution, adsorption to surfaces, ionization, hydrolysis, photolysis, microbial degradation, volatilization and partitioning (Baughman and Lassiter, 1978).

Stern and Walker (1978) stressed that studies designed to assess chemical fate in the environment can yield important information with regard to the area over which pollutants will be distributed, the chemical forms and concentrations in which the pollutants will be deposited, the types of reactions to which they will be subjected during transport and after deposition and the biological targets most likely to be sensitive to the pollutants' effects. Also, knowledge of expected fate is relevant to the design of the sequence of biological tests and conditions relative to the site of impact, concentration, period of exposure and specificity of biological effect. We note here that although acute and chronic toxicity test protocols have been widely standardized, standardization of fate assessment has lagged behind. Although studies of pollutant fate continue to receive increasing attention, this is an area that is ripe for improvement. We address this issue in more detail in Chapter 5.

2.7.3 Bioaccumulation

Recent schemes proposed for the classification and labelling of chemicals and chemical mixtures for environmental safety (Freij, 1991) emphasize three criteria:

1. acute toxicity to fish, *Daphnia* and algae;
2. bioaccumulation potential;
3. biodegradability.

Bioaccumulation refers to the uptake and retention of chemicals by organisms via food or water. The process by which organisms take up chemicals from water is defined as **bioconcentration**. Bioconcentration potential can be described by the ratio of the concentration in organisms to the concentration in water. At least for organic pollutants, the bioconcentration potential increases with decreasing pollutant solubility in water.

Uptake of chemicals from food and water may result in **biomagnification**, a term that describes the increase in tissue concentrations of a chemical as it passes up the food chain. The degree of biomagnification tends to increase with increasing pollutant half-life (Rand and Petrocelli, 1985). In practice, attempts to estimate biomagnification can be hindered owing to difficulties assigning organisms to particular trophic levels. It is generally assumed that organisms consumed as prey contain pollutant residues that are representative of the entire prey population, that these residues are at steady state and that emigration and immigration do not influence the measured concentrations (Moriarty, 1983). It is certainly plausible that organisms immobilized, or otherwise impaired, by unusually high body levels of pollutant will be preferentially consumed by predators. Also, individuals severely weakened by pollutant exposure, and therefore highly susceptible to predation, may take up increasing amounts of pollutants as their physiological systems begin to break down. Often biomagnification studies compare whole-body concentrations of pollutants in smaller organisms with individual organ concentrations in larger organisms from higher trophic levels. Such comparisons can be misleading since most pollutants are not uniformly distributed among body tissues and fluids. The measured concentration in one tissue can only be used to estimate the concentration in another if the organism is at steady state (Moriarty, 1983). Moriarty (1984) recommended the use of compartment models in favour of food chain models, because compartment models explicitly focus on the rates at which pollutants enter and leave organisms (or parts of organisms).

2.7.4 Biodegradation

The term **biodegradation** can be generally defined as 'any biologically-mediated process which results in the conversion of an organic chemical into organic and inorganic end products which are chemically distinct from the parent material' (Shimp *et al.*, 1990). Primary biodegradation, or **biotransformation**, has been defined as the 'loss of parent chemical identity', and ultimate biodegradation (mineralization) as the 'formation of microbial biomass and inorganic end products from a chemical.' As the breakdown products of chemicals that only undergo primary biodegradation may persist in the environment, estimating the complete mineralization of organic chemicals into harmless end products is of greater relevance for estimating potential environmental hazards. The biotransformation and degradation of most organic compounds present in the aquatic environment is largely accomplished by microbes (Rand and Petrocelli, 1985). Rates of biodegradation for a variety of organic chemicals have been described by a simple pseudo first-order model in which the biodegradation rate is directly proportional to chemical concentration (Shimp *et al.*, 1990). Such a model assumes that other fate processes are insignificant and that the chemical residence time (that is, the average time available for biodegradation to occur within a defined environmental compartment) provides a reasonable approximation of the transport processes in a given environmental compartment. The potential of a chemical to biodegrade can be estimated as the ratio between the biodegradability half-life (the time required to reduce the mass of a chemical to 50% of its original value) and the chemical residence time. Substances that have half-life/residence time ratios greater than about seven may be considered essentially nonbiodegradable (Shimp *et al.*, 1990).

It is well-known that many chemicals reaching the aquatic environment may be rapidly removed from the water column through adsorption and transport to bottom sediments, and that sediment-dwelling microbes play a critical role in the mineralization of organic substances. Unfortunately, biodegradation models for sediments are in a relatively primitive state of development. In Chapter 5, we address some of the deficiencies in this area and propose a number of suggestions for their improvement.

2.7.5 Uncertainty factors

It remains that much of the data used in risk assessment is derived from laboratory experiments of acute and chronic toxicity of a few selected species.

How do we use these values to predict risk under natural environmental conditions? In reference to avian and mammalian toxicity, Kenaga and Lamb (1981) stated that the difference in concentration between acute and chronic effects is usually between 10 and 1000-fold. Thus, if the environmental concentrations are estimated to be less than 1000 times the laboratory acute LC_{50}, Kenaga and Lamb (1981) considered it unlikely that toxicity would occur in nature. As we demonstrate in Chapter 4, it may not be possible to predict chronic from acute toxicities with an accuracy less than a few orders of magnitude. It appears that the closer we look (e.g., within single species or chemicals; Schimmel *et al.*, 1983; Clark *et al.*, 1989) the less we are able to make meaningful predictions between acute and chronic toxicity. Nevertheless, the relationship between acute and chronic toxicity is very often used to generate so-called safety factors for pollutants. Seager (1988) referred to the application of safety factors as 'little more than reasoned guesswork.' More appropriate terms, which have appeared in the literature, are '**application factor**' or '**uncertainty factor**'. Determining the size of the uncertainty factor depends on interspecies and intraspecies differences in toxicity, the completeness or adequacy of the data and the biological significance of the toxic effects (for example, greater safety factors are used if the effects are irreversible) (Kuiper-Goodman, 1989).

An explicit formula for the application factor (AF) has been given as the ratio of the **maximum acceptable toxicant concentration** (MATC) to the 96-hour LC_{50} for the same species (Ernst, 1980). The MATC can be determined in chronic toxicity tests (using either the entire life cycle of an organism or its most sensitive life stage). The MATC is sometimes regarded as equivalent to the observed 'no effect concentration' (Maki and Duthie, 1978). The American Institute of Biological Sciences defined the MATC as 'the highest toxicant concentration to which a test organism can be exposed without inducing a toxic response. In chronic toxicity tests it is typically described as a value lying between the highest observed "no effect" concentration and the lowest observed "effect concentration"' (American Institute of Biological Sciences, 1978). The AF is then used to calculate the MATC for other test species or test conditions for which LC_{50} tests, but not chronic tests, have been performed.

Stern and Walker (1978) cautioned that the following points should be noted with regard to MATCs.

1. Confidence in the calculated MATC is positively correlated with the relatedness of the species used.
2. The greater the margin between expected environmental concentration and calculated MATC, the more relaxed the acceptance of the calculation.
3. Confidence in the calculated MATC is increased when the LC_{50} values fit expectations and when there is an abundance of chronic data allowing validation of MATCs for the class of chemicals to which the test substance belongs.
4. Given the numerous modes of toxic action of chemicals there is little

biological justification for assuming that a reproducible ratio exists between an MATC and an acute LC_{50}.

The studies of Schimmel *et al.* (1983) and Clark *et al.* (1989), among others, provide data in support of this last point.

2.7.6 Hierarchical testing schemes

Risk assessment is often performed via an hierarchical series of tests designed to minimize the time and expense of superfluous testing while maximizing the accuracy of the assessment. Lloyd (1980) provided an example of the type of hierarchical sequence of steps that can be employed in risk assessment programmes. Stage 1 consists of acute screening tests on a few standard species. For some species (freshwater fish) and chemicals there is enough information on the MATCs so that the risk can be determined by the ratio of the MATC (or no-effect concentration, NOEC or NEC) to the maximum predicted environmental concentration. If this risk is determined to be acceptable, then no further tests are required. During stage 2, a series of predictive tests are performed to enable a more accurate estimate of the risks to be made. Lloyd (1980) stated that tests at this stage are selected for their simplicity and high predictive value. These tests are specifically designed to obtain more information on differences:

1. in sensitivity between the test species and the species for which protection is desired;
2. between the concentration causing acute toxicity and the minimum concentration causing harm to the organism;
3. in toxicity resulting from age or life stage;
4. in toxicity resulting from abiotic factors.

In stage 3, the validity of results from stage 2 tests is confirmed under field or simulated field conditions. If a large discrepancy is found between laboratory predictions and field observations, the cause of this discrepancy must be determined. Finally, in stage 4, environmental monitoring subsequent to the release of the chemical is performed.

It is generally recognized that the amount of information needed to make a sound hazard assessment will vary among chemicals. Hierarchical testing schemes commence with simple, inexpensive range-finding tests using single species and proceed through a stepped sequence of tests which increase in complexity, sophistication, cost and often duration (Cairns, 1981). The goal of such hierarchical testing schemes is to limit the time and expense of the most thorough testing to those substances of greatest potential danger. In most cases, this may be an efficient and effective strategy in that it permits large numbers of substances to be screened and it does not lead to burdensome and expensive testing of substances which very clearly pose no threat. However, for such a hierarchical scheme to be sufficiently protective, the criteria for proceeding (or not proceeding) from one step to the next must be scientifically sound. In the scheme presented above, a few relatively short-term high-dose studies on a few standard test species provide the first, and often only, test for most potentially dangerous substances. In this scheme, substances testing positive at stage 1, will require testing at stages 2–4. Because a negative result at stage 1 means that no further testing is required, this is the level at which mistakes in assessment can have the most serious consequences.

Cairns (1981) advocated the use of simultaneous, instead of hierarchical, testing because, in general, information derived from toxicity tests in the earlier steps is not demonstrably correlated with responses obtained in the latter steps. He argued that the ability to predict sublethal effects, as well as effects at more complex levels of organization, from single-species tests has not been compellingly demonstrated. Because cost is a major consideration and, as time is money, the delay of the more complex tests until after the short-term tests are completed can offset some of the supposed cost benefits of sequential testing. Cairns (1981) made the interesting observation that the sequential nature of hierarchical testing schemes reflects the historical development of hazard evaluation and the increasing recognition that more sophisticated data are needed to make a sound environmental decision. This, he argues 'does not mean that we should repeat this evolution of thinking in the hazard evaluation process.'

2.7.7 *Risk assessment in practice*

How are new substances evaluated for their potential to harm humans and the environment before their release or manufacture? In general terms, risk assessment protocols are designed to integrate estimates of potential (human and environmental) exposure with information about biological effects. An example of a testing scheme proposed by The American Society for Testing and Materials for evaluating hazard to nontarget aquatic organisms consists of a three phase approach (Maki and Duthie, 1978; Figure 2.4). During phase I, information is gathered on the physicochemical parameters of the substance along with estimates of usage, disposal and release patterns. Available data on structurally similar substances are considered. Under phase II, minimal acute toxicity testing is performed for substances likely to reach the aquatic environment in substantial quantities. These would normally include acute tests with a warm water fish species such as the bluegill, a cold water fish species, for example, the rainbow trout, and *Daphnia* species (Kimerle *et al.*, 1978). Biodegradation potential and partition coefficients between water and solvents or solids are also estimated during this phase. Those chemicals requiring testing under phase III, would undergo lifecycle chronic testing and tests for bioconcentration. Ultimately, the assessment of risk is based on the toxicity of the test substance and on its estimated environmental concentration.

From the industrial side, Branson (1978) emphasized the importance of

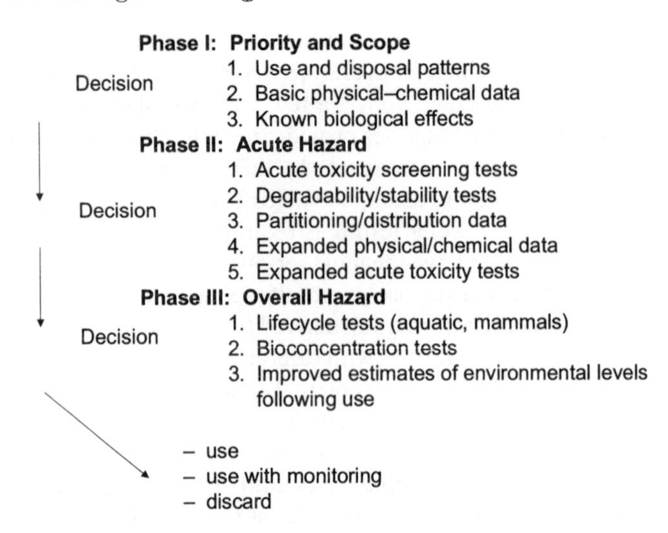

Figure 2.4 Example of an hierarchical testing scheme proposed for evaluating the hazard of chemicals to aquatic ecosystems. Adapted from Maki and Duthie (1978).

being able to rank chemicals early in the development stage so that efforts can be focused on those that are safest. Brungs and Mount (1978) cautioned on the need to account for accidents and carelessness in assessing potential impacts. This would require information on the mode and amount of material transported as well as other information on handling and human contact with the material.

The Commission of the European Communities (CEC) requires that the manufacturers of new substances provide notification, for use in classification, packaging and labelling, before the chemicals are approved for sale. A base data set is required for all substances. It includes basic information about the substance, such as its name, molecular structure, purity, amount and kinds of additives, spectral data, methods of detection, information on the proposed usage of the product (for example, for what purpose, by whom, in what amounts produced or imported), various recommendations on handling, storage, transport and emergency measures. Physicochemical properties of the substance must be provided. These include its nature (colour and physical state), melting and boiling temperatures, relative density, vapour pressure, surface tension, water and fat solubilities, *n*-octanol/water partition coefficient, flash point, flammability, explosive properties, autoflammability, oxidizing properties and any other available information. Toxicological effects are estimated from tests of acute toxicity (administered orally, by inhalation and cutaneously), acute skin and eye irritation, skin sensitization, subacute toxicity (over 28 days) and mutagenicity (in bacteria, non-bacterial *in vitro*, non-bacteria *in vivo*). Required ecotoxicological information is limited to tests of acute toxicity in fish, acute toxicity in *Daphnia*, biodegradability and hydrolysis as a function of pH. Information is also provided regarding the possibility of rendering the substance harmless, for example, through recovery or recycling, neutralization or destruction.

2.7.8 Obstacles to effective risk assessment

What factors limit our ability to effectively assess realized and potential pollutant effects? Uncertainties in the risk assessment process include gaps in available data as well as uncertainties in the validity of model assumptions (Hart and Jensen, 1992). The exact chemical identification of individual substances present under field conditions is often difficult (Kinne, 1980). Certain substances, such as pesticides, released into the environment result in complex mixtures whose properties and ecological consequences are difficult to determine. Thus, Lewis (1980) noted that most measures designed to reduce pollution have focused on limiting effluent inputs rather than on assessing and responding to responses *in situ* of populations and communities. Numerous processes (chemical, physical and biological) influence, modify or mask pollutant identity and effects. For example, the toxicity of substances such as PCBs differs from that of metals in that the toxic effect of the former depends on the structure of the intact molecule so that when it is broken up into its constituent elements the biphenyl is no longer toxic (Moriarty, 1983).

In addition to complications arising from biological or environmental variables, the employment of arbitrary decision criteria continues to impede our ability to accurately assess the risks of pollutant exposure. In many instances, standards and protocols seem to arise from committee or workshop discussions. Rarely in governmental directives or regulatory guidelines are justifications provided for the exact decision criteria. For example, an acute LC_{50} value of 1.0 mgl^{-1} is often cited as a critical value for determining the toxicity of chemicals to aquatic organisms (Maki and Duthie, 1978; Organization for Economic Cooperation and Development, 1989; Nordic Council of Ministers, 1990 and others). In none of these reports are any references cited to justify the choice of 1.0 mgl^{-1}. Why not 0.9 or 1.1? Is one decimal place necessary? Is it sufficient? At what level of statistical significance (and power) are LC_{50} values judged to be different from 1.0 mgl^{-1}? The AIBS can be commended for at least admitting that their choice of values is arbitrary. They stated, 'In those situations where no chronic toxicity data has been generated and an experimentally derived application factor is not available, an *arbitrary* value of 0.01 has been used for those chemicals suspected of possessing substantial chronic toxicity and a value of 0.1 has been used for those chemicals believed to have minimal chronic toxicity' (American Institute of Biological Sciences, 1978).

When hazard criteria are decided by group consensus, the implication is that these decisions are made by a select body of persons having great expertise in the field. This may well be true. But strange things can happen in committees, and numbers selected by general agreement to provide rough guidelines for decision making have a way of taking on magical qualities. Sometimes they even become law. For example, the North Sea Convention decided to set reductions in pollutant and nutrient additions at 50%. When asked why 50% was chosen as the target value, one negotiator responded, "It's an easy number to remember." And so 50% it was (J. Gray, personal communication).

A number of industrial scientists with whom we spoke, understandably expressed great frustration at having to meet such stringent 'guidelines' set by the regulatory agencies (an LC_{50} of 1.06 mg l^{-1} just would not pass muster). Perhaps 1.0 mgl^{-1} is a magic number, but this has yet to be adequately proven (and subsequently cited) in the ecotoxicological literature. All would agree that guidelines are valuable and that enforced regulations are necessary to achieve environmental protection. Obviously, if a regulation is set at a specific value, it is critical that such standards be strictly enforced. Thus it is essential that guidelines and laws, which are purported to be scientifically justifiable, be based not solely on group consensus but on hard data that is thorough, robust, relevant, and has been subjected to peer evaluation.

2.8 Decisions in risk management

2.8.1 Objectives and limitations

Truhaut (1977) maintained that 'As in human toxicology, the ultimate objective of research in ecotoxicology should be the establishment of protective measures against the harmful effects of environmental pollutants on the various constituents of ecosystems.' This is the primary objective of environmental managers. Environmental managers make decisions for environmental protection based, in part, on data gathered in risk assessment analyses, the collection of which, ideally, is based on theoretical and empirical principles developed in ecotoxicological studies. The specific risk management strategy more or less dictates the type and detail of the risk assessment. For example, procedures regulating the classification and labelling of chemicals require only that the hazard identification step be performed (Hart and Jensen, 1991).

Management decisions must often be made in the absence of a good understanding of the system of concern. As Anderson (1989) noted, we should recognize that in such cases management itself is usually an experiment. Although underused as a research tool, the deliberate adoption of various management strategies can provide useful comparative data. When the requirements or consequences of different activities (such as among different fisheries, or between fishing and industry) come into conflict, managers must determine which activities are to be permitted and which are to be prohibited. In the mid to late 1970s, there was intense debate concerning the development of offshore oil wells in the New York Bight (Squires, 1983). Such plans were in conflict with the needs of the coastal fisheries since oil spills, noise and other disturbances associated with the drilling were feared to harm or drive away the local fish (effects that could potentially outlive an oil field's lifetime). It was argued that the recreational industry surrounding the Bight and the fisheries provided an economically more important and more long-lasting resource than could be provided by short-term (about 20 years) exploitation of oil reserves.

Whereas detection of pollutant effects is exclusively a scientific problem, prevention and treatment of pollutant effects requires consideration of political, economic and social factors in addition to scientific understanding. However, deciding to what extent each of these factors should contribute to environmental decision making is no trivial matter. 'It may be expected that in the absence of management, mainly short-term economic advantages and short-sighted political arguments will decide the course of events' (Wolff and Zijlstra, 1980).

> Wherever we have looked in the ocean, the twin evils of greed and neglect were present. Greed has led to over-harvesting of the living resources: the catching of fish beyond the capability of a stock to renew itself. Neglect has usually shown itself by using the ocean for disposal of wastes cheaply, a shortsighted process which may have long lasting effects on other resources (Squires, 1983).

Once we have identified and evaluated the risks involved with various human activities, we face the challenge of managing our activities in a way that permits reasonable use with the minimum harm to the environment. Lewis (1980) expressed concern over the deployment of a series of uncoordinated responses based on local economic needs or opportunities and argued for a coherent national management plan. Achieving effective management strategies is complicated by the fact that the responsibility for managing the environment falls on a great diversity of cultural, political and judicial groups of various sizes (some of which overlap) that have no relation to natural ecological boundaries or to the distribution of the biota that we wish to protect.

In the USA (Squires, 1983), Australia (Hutchings, 1991) and the UK (Elliot, 1991) the management of the coastal zone appears to be left to a number of organizations or jurisdictions, each of which is responsible for particular facets of coastal planning and regulation, and each of which is primarily concerned with protecting its own interests. A recent UK report called for a unified coastal zone management plan (Gubbay, 1990, reviewed by Elliot, 1991). In his review Elliot wrote, 'Many would consider that the call for a single [plan for coastal zone management] is impractical as it would have to rectify the many built-in conflicts of competing interests which could make it impotent.' He doubted whether such a unit could determine whether recreation, conservation and wildlife interests should have precedence over commercial and development interests and further suggested that instead of centralization, what is needed is simply a more well-defined policy. He concluded that the key to the coastal zone management problem is 'greater coordination and the ability to resolve conflicts of interest without producing excessive over-regulation.' To this we would add that effective long-term management is achievable only to the extent that it is based on geographic rather than political boundaries. As our effects on the environment can extend to a global scale, it is becoming increasingly obvious that national and political boundaries do little to facilitate solutions to environmental problems. A pertinent example is presented in a recent editorial on climate change (Gray, 1991). Although firm proof may be as much as 10 years away, scientists are generally agreed that there are strong reasons to believe that an artificial climate change is occurring. International political responses to threats of climate change have been predictable and ineffectual. As of 1991, the USA and the former Soviet Union produced the highest percentage of the world's carbon dioxide discharge, 22% and 18.4% respectively. According to Gray (1991), the official USA policy on climate change is apparently to 'do nothing until concrete proof is obtained that man-made changes are occurring.' In contrast Norway, which produces a mere 0.22% of the world's carbon dioxide has agreed by the year 2000 to stabilize emissions at 1989 levels; an action that will have a negligible effect on global production but which will impose a heavy economic burden on Norwegians.

Economic constraints are most often claimed in defence of failures to establish environmentally clean practices. For example, the cost of recycling has been deemed prohibitive in many instances. In poorer countries, economic constraints may sometimes justify the adoption of environmentally inferior practices; however, even affluent countries make use of the economics argument. For example, the USA and Canada justify the removal of the last remnants of ancient forests on the grounds of economic necessity (Soulé, 1991). Such actions strongly suggest that the true cost of the destruction of these forests has not been accurately quantified. Costanza *et al.* (1991) noted that when resource depletion and degradation are factored into economic calculations, a radically different picture emerges from that depicted by conventional standards, such as gross national product. The danger is that if we continue to ignore the value of natural ecosystems, we may actually drive the economy down while we think we are building it up.

As the geographical scale of our environmental problems increases, the political, cultural and social barriers to their effective solution become more apparent. The danger in any such problem solving activity is that the compromise solution, to which all deciding parties can reach an agreement, may result in solutions that are politically fashionable but environmentally ineffective. Lindzen (1990) claimed, 'It is easy to see that every suggested policy designed to prevent [global] warming will have almost no effect on warming.' In general, we can expect that the bigger and more complex the problem and the more parties involved, the more challenging it will be to devise solutions that are both mutually acceptable and effectual.

2.9 Summary and conclusions

Protection of ecosystems is the explicit goal of much current legislation related to the environment. It is thus implicitly assumed that the essential properties of ecosystems can be identified and measured. Recent debates in ecology suggest that this assumption has not been fully confirmed. Complications arising from the spatial and temporal complexity of ecosystems impede the detection of pollutant effects. There is presently no general theoretical relationship that links pollutant effects on individual species to effects on intact ecosystem structure and function. Whether such a theory or family of theories is even possible remains to be determined.

At the ecosystem level, concern has often centred on the effects of pollutants on ecosystem 'health' or 'well-being'. Although clear and generally consistent changes in ecosystem properties have been shown to occur in response to stress (including pollutant stress), the analogy between such properties and the term 'health' as it applies to individual organisms is obscure. We suggest that such terms are inappropriate because they presuppose that objective measures of ecosystem health or well-being exist and because they do not accurately reflect the important role of value judgments in ecosystem protection and management.

One of the greatest difficulties in achieving successful environmental protection is persuading society to balance relatively abstract or immeasurable qualities of ecosystems against concrete economic figures. Environmental scientists need to develop more convincing arguments for protecting ecosystem structure and function that can withstand society's natural tendency to seek short-term economic and political gains. Although science can be used to determine whether significant effects of pollution on natural systems occur, society must judge which effects are unacceptable. Because environmental management is not only a scientific issue, it is critical that the public is adequately informed. Historically, the mass media has played an important role in public awareness of environmental issues. Experience has shown that this is an ineffective mode of public education. There is a pressing need for more rational long-term approaches to resource and environmental management. This will require changing the public's impression that the environment is an intellectual issue which has little bearing on daily individual behaviour.

Experience with open access or multiple-use resources has shown that active management is essential to maintaining long-term sustainability. Environmental management, as it is currently practised, involves much guesswork. As environmental problems increase in number and scale, so does the need for effective and scientifically sound management strategies at the international level.

3 Measuring ecotoxicological effects on populations, communities and ecosystems

... there is no single biological measurement that will serve to indicate the effects of pollution ... The pressures to adopt biological measurements within environmental monitoring programmes are great and, as a consequence, so are the temptations to expect (and to claim) more of the available biological techniques than is scientifically reasonable.
Brian Bayne and others in an overview of the Group of Experts on the Effects of Pollution (GEEP) workshop, 1988

While it has become abundantly clear that there is no single universal test or bioindicator for the detection and quantification of pollution, there is even a lack of consensus as to which approaches may permit accurate, efficient and predictive assessment of pollutant impact. Ecotoxicology is at a critical turning point. Ecotoxicologists have demonstrated that pollutants exert effects at all levels of biological organization, but we are still unsure of how effects at different levels of organization relate to each other. Stripped of the relatively recent context of pollutants, we recognize an old biological problem. In the past the tendency has been to circumvent the problem of an interacting hierarchy by defining separate fields of study corresponding to each biological level. Ecotoxicology cannot afford this luxury. We must address the hierarchical problem directly. To make progress we will need to draw heavily from disciplines that have grappled with questions of hierarchy in the past. Prime examples are the fields of genetics, physiology and ecology. Given the difficulty of the task, it is essential that advances in these closely allied disciplines be used to the full.

In Chapter 1, we introduced the general categories of tests for detecting the effects of pollutants on biological systems and distinguished between observational and experimental test designs. In this chapter we examine in more detail several methods that have been employed to study ecotoxicological effects in the field and laboratory and at different levels of biological organization. We begin with a look at the challenges involved in designing and conducting field monitoring programmes. Next, approaches for measuring pollutant effects at the community level are analysed. From here, we move to a discussion of techniques developed for examining the responses of populations to pollutants. We have divided the population-level approaches into those dealing with whole organisms, those dealing with physiological and biochemical measures of organism response and those focusing on population genetics. We attempt neither to provide an exhaustive list of every ecotoxicological test ever developed nor to present a detailed recipe for carrying out specific tests. For the latter, the reader is directed to the appropriate primary literature. Our intention is to provide a general description and critique of current approaches for studying pollutant effects on living systems in order to highlight promising directions and areas in need of improvement. By focusing on the advantages and limitations of selected methods, we hope to draw attention to the kinds of problems that can complicate the detection and interpretation of pollutant effects at different levels of biological organization. Although all of the methods we discuss in this chapter have been applied to the study of chemical pollutants, we emphasize that most of them have not been employed for regulatory or management purposes. Some of the methods are in a preliminary stage of development, whereas others have, for various reasons, been deemed inappropriate for regulatory use. This by no means invalidates these methods, but it highlights the difference between the needs of ecotoxicological research and those of regulatory testing. We return to this point in Chapter 6 and conclude that it is the failure to recognize this distinction that has led to much unproductive debate and haggling over the value of various ecotoxicological tools.

3.1 Field monitoring

A well-designed monitoring programme can reveal important information about the rates of release of pollutants into the environment, about the degree of and changes in environmental contamination and about the biological effects of pollution (Green, 1979). Monitoring of pollutant concentrations in biotic and abiotic components of ecosystems, as well as of pollutant effects on various biological structures or functions have been employed in monitoring programmes.

The goals of most pollution monitoring programmes used to date have been to determine the average level of pollution in an area and its change over time and to identify point sources of pollutant input. Several factors limit the effectiveness of pollution monitoring programmes. First, it is often impractical to measure all contaminants, and the sensitivity of routine analytical techniques may limit the types and levels of pollutants that can be measured. Second, regularly scheduled sampling may result in occasionally high, and biologically significant, values being missed. Third, the biological significance of the pollutant levels found may not be clear and, if more than one pollutant is present, interactions can complicate interpretation of the results.

Given the impracticality of monitoring all or most pollutants and species, great efforts have been devoted towards the selection of appropriate indicator organisms which could provide a convenient index of the extent of pollutant damage. Several studies conducted in the early 1970s found a significant relationship between pollution level and species composition of marine benthic communities. Pearson (1970; 1971; 1975) investigated changes in benthic species composition in response to the release of effluent from a paper and pulp mill on the west coast of Scotland. Rosenberg (1972; 1973) examined the recovery of the benthic community in a Swedish fjord following the closing of a wood pulp mill. On the east coast of the USA, Sanders *et al.* (1972) investigated the effect of an oil spill on the benthic faunal community. All of these studies found significant changes in species composition in response to organic pollution and dominance by a very few species (mostly annelid worms) under the most polluted conditions. The presence of these so-called pollution indicator species was suggested as offering a potentially useful measure of the degree of pollution in an area. However, Pearson and Rosenberg (1978) questioned the use of presence or absence of certain characteristic species as evidence of pollutant impact. Their work demonstrated that the ecological succession occurring in response to a pollution gradient is complex and dependent on a number of variables in addition to the pollutant concentration. Nevertheless, consistent and significant changes in community structure have been shown to occur in response to pollution. We discuss these changes, and methods that have been used to study them, in section 3.2.

Also in the early 1970s, it was recognized that measurements of pollutant residues in aquatic organisms could provide a valuable addition to pollutant analyses of water and sediments (Phillips, 1980). The idea behind such biomonitoring approaches is to use the organism's pollutant load as an index of exposure. According to Phillips (1980), the following are the advantages of measuring pollutant residues in aquatic organisms.

1. The concentration of many pollutants is higher than in the surrounding environment.
2. Only the fraction of pollutant that is biologically available is measured.
3. If the rates of uptake and excretion are known, a time averaged index of pollution is obtained.

Biomonitoring has been found to be quite useful for pollutants such as heavy metals and organochlorine compounds. However, it is well recognized that both biotic and abiotic variables influence pollutant kinetics and must be accounted for in interpreting biomonitoring data (Phillips, 1980). Although good correlations between pollution by some metals and metal content in some species have been found (Phillips, 1977), the interpretation of metal body burdens may not be as simple as originally hoped. Depledge and Rainbow (1990) emphasized that the handling of metals within an organism and the physiological condition of the organism can determine the significance of a particular body load of metal. For example, differences in the partitioning of metal among the different tissues in an organism may have a large influence on toxicity, but may be masked when measurements consist of whole body metal load. Whereas total metal load may indicate pollution hot spots, Depledge and Rainbow (1990) argued that it is not sufficient for assessing or predicting biological effects. They cautioned that, until a greater understanding of trace metal handling by organisms emerges, monitoring metal loads in organisms may not be a suitable approach for mapping damage due to metal pollution.

Very few species meet all, or even most, of the criteria for an ideal bioindicator species. Bivalve molluscs appear to be one of the most suitable groups in that a number of species are:

1. relatively tolerant to pollutant exposure;
2. sedentary;
3. abundant;
4. relatively long-lived;
5. of a large enough size so that tissue analyses can be made;
6. easy to sample and relatively tolerant of laboratory conditions;
7. euryhaline;
8. do not appear to regulate their body loads of at least some pollutants (Phillips, 1980).

Nevertheless, even when good bioindicator species are identified, there still remains the significant problem of linking measured residues to biological effects.

Species selected for biomonitoring purposes have generally differed from those used in standardized laboratory toxicity tests. The latter are often selected to be among the most sensitive species to chemical effects whereas the former must be able to tolerate a relatively wide range of polluted conditions. This discrepancy has done little to facilitate the linkage between laboratory test results and pollutant impact in nature. Also, it is now well established that organisms vary widely in their sensitivity to different chemicals, that no single species is the most sensitive to all toxicants, and that the probability of selecting the most sensitive species for testing a particular chemical is rather small (Sloof *et al.*, 1983; Blanck *et al.*, 1984; Cairns and Smith, 1989). Because species that are widely distributed are, by definition, tolerant of a range of environmental conditions, they are probably less sensitive than many species in their tolerance to pollutants. However, there is no reason to suspect that their tolerance, relative to other members of their community, is constant among different locations and for different chemicals. Klerks (1987) noted that the validity of the pollution indicator concept depends on the absence of differences in resistance between different populations of a species. He found genetic differences in cadmium resistance among populations of an estuarine oligochaete, and related these differences to variation in metal uptake (in this case, the more tolerant population took up more metal).

A further limitation of residue biomonitoring is that equating body load with exposure or biological effects is not valid for pollutants that do not bioaccumulate. Regular monitoring of pollutant residues in the environment or biota is not likely to be a useful strategy for assessing potential hazards associated with certain highly toxic chemicals, as significant biological effects can occur in the absence of measurable residues. For example, some of the pyrethroid insecticides have been found to cause substantial acute mortality in estuarine populations at concentrations well below the detection limit (Schimmel *et al.*, 1983; Clark *et al.*, 1989). Monitoring of water, sediment and biota from estuaries in South Carolina, USA indicated little potential impact to estuarine fauna as a result of insecticide exposure (Trim and Marcus, 1990). Despite this, 30 insecticide-related fish kills, representing 11.5% of all estuarine kills, were reported during this monitoring period. Trim and Marcus (1990) concluded that monitoring data may not be useful for providing early detection of pesticide impact or for identification of pollutant sources. The following reasons were outlined.

1. Monitoring stations are usually situated in larger water systems, whereas acute pesticide impacts are typically site-specific events of limited area, such as in tributaries or small coves close to shore.
2. All potentially relevant pesticides may not be measured.
3. Sampling schedules may not be extensive enough to detect pulsed or intermittent inputs.

In some situations it may be efficient to monitor chemical residues in organisms, rather than to attempt to directly measure biological effects. However, when effects occur at pollutant concentrations below our ability to chemically detect them, direct studies of biological effects are necessary to assess impact. Because a major goal of monitoring studies is to obtain an early warning of the effects of stressors, Gray (1989) suggested that detection should focus on impairment of individual performance. Underwood and Peterson (1988) argued, however, that reductions in individual performance (growth or reproduction) may not necessarily have any impact on population dynamics or on the contribution of populations to food webs for those species, for example, many marine invertebrates, in which reproductive rates are decoupled from recruitment to adult populations. Nevertheless, great efforts have been directed toward identifying pollutant effects at the suborganismal level (that is, molecular, biochemical or physiological changes) and towards developing these so-called biomarkers to the degree that they can be incorporated into routine monitoring programmes. The term biomarker is rather general in that it appears to include any and all pollutant effects occurring at the organism level or below. The advantage of biomarkers over bioresidue analysis is that effects on biological structure or function are measured directly. Biomarkers that respond to a wide variety of stresses, such as mixed-function oxidases, and to very specific stresses (metallothioneins, for example) have been identified. To date, none of these have been developed to the extent that they are ready to be widely incorporated into risk assessment protocols; however, a number of them show promise. The biggest challenge in the future development of biomarkers is to demonstrate that changes in structure or function at the suborganismal level are linked to changes at the population and community level. The list of potentially useful biomarkers currently under investigation is extensive. We review a few of them in section 3.4 and direct the reader to recently published books on the subject for more thorough coverage (Peakall, 1992; McCarthy and Shugart, 1990).

3.2 Communities

3.2.1 Diversity indices

Crow and Taub (1979) criticized single-species toxicity tests as inadequate because they measure only 'the degree to which a pollutant prevents an organism from reaching its biological potential of life span and reproduction, if all other conditions approach the optimal.' In nature, all other conditions are rarely optimal due to competition, predation, disease, physical factors etc. Measuring responses to pollution at the community level may overcome some of the difficulties involved in extrapolating from effects on one or a few species to entire assemblages.

Reductions in species diversity, retrogression to opportunist species and shifts to smaller-sized species are all well-documented structural responses of communities to stress (Gray, 1989). Diversity indices are measures designed to integrate the numbers of species and their relative abundances in a community to provide an estimate of community complexity (Gray, 1980). They are often used to determine whether a community has been affected by pollutants, with a drop in diversity indicating significant pollutant damage. Gray (1980) summarized the many limitations involved in the interpretation of diversity indices for pollution studies. He cited evidence that, in addition to pollution, predation and competition, spatial heterogeneity and successional stage have been shown to alter community diversity. As we discuss below, diversity indices have received a great deal of criticism, particularly in terms of their use as pollution indicators. However, no clearly preferred alternative for analysing changes in community structure currently exists.

In general, biotic indices can only be devised if a great deal is already known about the ecosystem and the pollutant. The quality of the measure may be largely dependent on accurate identification to the species level. As a number of species typical of polluted habitats have been shown to be sibling species (e.g., Grassle and Grassle, 1974), this can make accurate identification of large field samples technically impractical. However, Gray (1979) suggested that correct identification to the sibling species level, or even to the species level, may be unnecessary for an overall detection of pollutant effects.

Diversity indices generally include a limited subset of species in the community and do not take into account the relative importance or ecological function of different species. Such indices may be unaffected if one species replaces another, and are unlikely to be affected if rare species disappear. To understand the effect of the pollutant, changes in diversity indices need to be related to changes in the dynamics of individual populations. In addition, comparisons of diversity among communities should be restricted to roughly comparable groups of species (Moriarty, 1983).

Green (1979) provided an overview of the statistical and biological problems associated with several common diversity indices. He noted that diversity is a vague concept which combines two different and often independently varying components. He cautioned that there may be serious problems in assuming that diversity, as calculated from a sample, is an unbiased estimate of the diversity of the community. Furthermore, he argued that there is little evidence that diversity can provide a general measure of environmental quality. Factors unrelated to stress, such as the age of the system, can have a large influence on diversity (Rapport, 1989). Green (1979) argued that 'other statistical methods retain more of the information in the biological data while reducing them to a more useful and ecologically meaningful form.' He provided examples of a number of multivariate approaches that, designed with specific and well-defined questions in mind, may provide more effective approaches to the analysis of pollutant effects on community structure. Williamson (1987) also recommended the use of multivariate techniques, such as principle component analysis, because such approaches do not equate all species and because they are sensitive to the correlation structure (that is, interactions) within the community. He added that measures such as species number and total numbers of individuals, as well as diversity indices, are insufficient because such measures ignore the particular species involved in community change. An example of the use of multivariate analyses for community pollution studies is provided by Gray *et al.* (1990).

3.2.2 Other measures of community structure

Gray and Pearson (1982) analysed changes in species structure resulting from several stressors including organic enrichment, oil and titanium dioxide. They found that in general, the first response to the stress was a decrease or elimination of some of the rarer species in the community and an increase in abundance of some of the moderately common species. If the stress was severe, many rare species were eliminated and a few opportunistic species became dominant. In addition to changes in species abundance, a trend toward smaller-sized species has been shown to occur in response to stress from oil, copper and organic enrichment in aquatic habitats and irradiation and burning in terrestrial habitats (Gray, 1989). This decrease in size is related to an increase in the proportion of opportunistic species.

Gray (1979) suggested that deviations from the log-normal distribution of individuals among species could provide a more sensitive indication of pollutant impact than more traditional indices of species diversity. The log-normal distribution commonly characterizes large, heterogeneous samples as a result of many independent factors acting multiplicatively (Gray, 1980). He noted that departures from the log-normal distribution can occur in response to a variety of environmental disturbances which alter the equilibrium community structure (assuming that an equilibrium exists, though it may not; e.g., Connell and Sousa, 1983; Williamson, 1987). For example, seasonal recruitment of juveniles can cause the community to temporarily depart from a log-normal distribution. For this reason, Gray (1980) recommended that monitoring year-to-year variations in species abundances would be most efficiently conducted during the winter in temperate habitats, or at those times of year when populations are not actively recruiting.

Fitting the log-normal distribution to field data requires that a large number of heterogeneous species (that is, from different taxa) are sampled and generally requires accurate identification to the species level. An advantage of using the log-normal distribution is that deviations can be tested statistically against the theoretical distribution. In contrast, diversity indices, such as the Shannon–Wiener index, that have no expected distribution, must be tested against a reference site or time.

Warwick (1986) and Warwick *et al.* (1987) proposed a graphical index of pollution-induced disturbance which is based on a comparison of the distribution of numbers of individuals among species with the distribution of biomass among species. This index is based on the idea that the effects of pollution on species abundance and biomass are essentially the reverse of those occurring during the process of ecological succession. McManus and Pauly (1990) adapted Warwick's graphical approach into numerical indices. These indices have been shown to provide useful indicators of gross pollution due to organic enrichment. Determining whether this approach is also useful for detecting low levels of pollution or pollution from other toxic chemicals will require more thorough testing.

Gray *et al.* (1988) compared a number of measures of community structure along a pollution gradient in the Frierfjord/Langesundfjord, Norway. Measurements of species number, number of individuals per species and biomass provided the basis for several univariate and multivariate analyses of pollutant effect. These measures were correlated with measured pollutant load at each

Table 3.1 Response of various measures of community structure to disturbance. Sites increase in water depth in the order E (22 m) < G (23 m) < A (76 m) < D (101 m) < C (108 m) < B (113 m). Site A has the lowest pollutant loading and site G has the highest

	Sites					
Community attributes	A	B	C	D	E	G
Total species numbers	—	—	—	—	—	?
Total abundance	—	—	—	—	—	—
Total biomass	—	?	Y	Y	—	—
Biomass/abundance	—	?	Y	?	—	—
Abundance/species numbers	—	?	Y	—	?	—
Faunal group abundance	—	Y	Y	?	—	—
Faunal group biomass	—	Y	?	—	—	—
Abundance of dominants	—	Y	Y	Y	—	—
Biomass of dominants	—	Y	?	—	—	—
Diversity and evenness	—	?	?	—	—	?
k-Dominance curves	—	?	Y	Y		—
Trophic group abundance	—	?	?	—	—	—
Individuals among species distribution	—	?	Y	?	Y	—

Y = clear disturbance; ? = possible disturbance;— = no disturbance.
From Gray *et al.* (1988).

field sampling site. Various descriptors of community structure differed in their apparent sensitivity to pollutant-caused disturbance (Table 3.1). Based on the summed indications of disturbance, three sites (C > B > D) appeared to be clearly disturbed. Total species numbers, total abundance, trophic group abundance, diversity and evenness were generally insensitive measures of community structural change. The abundance of dominants clearly indicated disturbance at three sites. Two of the three sites were indicated as clearly disturbed by measures of *k*-dominance curves, total biomass and faunal group biomass, whereas the third site (not always ranked in the same order) was determined to be possibly disturbed. Although the concentration of polycyclic aromatic hydrocarbons and several metals varied significantly among sites, the effects of these pollutants appeared to be masked by the effect of water depth as the overriding factor controlling community structure. Those sites that appeared to be disturbed (B, C and D) were located at depths of 101–113 m. Seasonal anoxia at the deep sites was suggested as the probable cause of disturbance, as there was no clear relationship between the various disturbance indicators and measured pollutants. Sites A (76 m), E (22 m) and G (23 m) did not appear to be seriously disturbed even though site G was closest to the pollution source. The site with the lowest pollutant loading was very similar in most of the biological analyses to the site having the highest pollutant loading, although several species usually indicative of organic loading were more abundant at the latter.

The authors concluded that several of the more classical methods, such as diversity indices or abundance of pollution indicator species, although probably quite powerful when applied by 'knowledgeable experts' are rather subjective and not suitable for formal statistical analysis. More objective analyses, such as the multivariate ordination techniques used in the study by Gray *et al.* (1988), may prove superior for situations in which adequate subjective experience does not exist. Unfortunately, the application of these multivariate techniques to examine pollutant effects in mesocosm experiments failed to detect clear differences among treatments (Gray *et al.*, 1988).

3.2.3 Manipulative field experiments

Underwood (1989) recommended using manipulative field experiments to measure the attributes of populations (inertia, resilience, stability) in nature. For example, he suggested that the kinds and levels of stress under which a population is able to maintain itself could be investigated experimentally by removing certain fractions of the population. This would permit predictions to be made of the response of the population to real stresses. In aquatic environments, manipulative experiments using artificial substrates can eliminate natural substrate heterogeneity. They have the advantage of selectively sampling the larval or juvenile stages of populations, which may be the most toxicant-sensitive members of the population and which may provide a better estimate of future population size than adults (Green, 1979). The disadvantages of artificial substrates include the fact that they may not accurately sample some species and that their sampling effectiveness may be very dependent on distance from the adult population and on accidents of fate.

Cairns *et al.*, (1986) designed a test system based on the colonization success of microbial communities on artificial substrates. This system measured changes in community structure resulting from species–specific mortality; reproductive impairment; and toxicant effects on interspecific competition and succession. The authors found that low levels of cadmium, under 10 $\mu g\ l^{-1}$, had no effect on the structure of an established community (i.e. that community acting as a source of potential colonists), but reduced the number of species colonizing the artificial substrates. Concentrations of cadmium nearly two orders of magnitude higher (500 $\mu g\ l^{-1}$), severely affected the structure of the source community as well as the colonizers. In contrast, both source community and colonizers were affected by the same concentrations (10 $mg\ l^{-1}$) of 3-trifluoromethyl-4-nitrophenyl (TFM) tested.

One of the drawbacks of this kind of test is that determination of microbial community structure requires fairly sophisticated training. An advantage is that, in one sense, field corroboration of laboratory predictions is more straightforward and simpler to perform than for some other techniques, because the same artificial substrates can be used in both field and laboratory. A similar type of approach has been employed in studies of benthic macrofaunal recruitment into sediments contaminated with different concentrations of pesticides (Tagatz and Ivey, 1981; Tagatz *et al.*, 1987). Some differences were found between the number and kind of species colonizing laboratory microcosms (inoculated by continuously flowing unfiltered seawater) and those colonizing sediment cores incubated in the field. However, both field and laboratory systems detected significant changes in community structure in response to very low levels (0.1–10 μg l^{-1}) of the pyrethroid insecticide, fenvalerate.

3.2.4 Microcosms

Microcosms overcome many of the problems of covariance associated with field monitoring studies and offer a higher level of complexity than single-species toxicity tests. There are both laboratory and *in-situ* models that vary in design and limitations. They may attempt to include most of the biota from the natural system or they may specifically select a limited subset of species usually representing different trophic levels. The latter, referred to as gnotobiotic systems, are artificial assemblages of organisms and, although having a greater degree of complexity than single-species tests, may not be expected to closely model ecosystem response (Crane, 1990). Field microcosms have proven very useful for studying the fate and chemical form of pollutants under natural conditions (Davies and Gamble, 1979). Such studies have revealed that many pollutants are rapidly lost from the water column and have thereby highlighted the importance of the benthic community as a likely target of pollutant impact and as an important mediator of pollutant degradation. Microcosms have been described as 'a bridge between standard laboratory effects tests and field studies that can account for much of the complexity and stability typical of natural environments' (Giddings and Eddlemon, 1979). Koehl (1989) pointed out that the use of microcosms facilitates the inclusion of adequate replicates and appropriate controls in ecological experiments and, unlike natural field experiments, microcosms can be designed to reproduce the assumptions of a model being tested. One of the (generally untested) assumptions involved in the use of single-species tests is that the economic savings gained by this approach exceed the cost of a poor management decision (Crane, 1990). However, Cairns *et al.* (1986) demonstrated the feasibility of designing highly cost-effective multispecies tests, particularly for microbial species.

Davies and Gamble (1979) considered several advantages and disadvantages of the microcosm approach. Microcosms are useful for ensuring that the same populations are sampled over time and that replicate systems can be experimentally manipulated. Because at least three trophic levels are frequently included, such systems are often self-sustaining for extended periods of time. Disadvantages of many microcosm designs for aquatic systems are that scale limitations (for example, long-distance dispersal is precluded) and changes in physical dynamic processes (such as reductions in vertical mixing and thus the fluxes of living and non-living components of the system) can limit data interpretation. Sometimes stirring devices have been installed to alleviate the latter problem. Although larger microcosms have certain benefits, for minimizing wall effects, for example, economic and technical constraints make it more difficult to replicate large systems.

As Cairns and Buikema (1984) noted, 'Microcosms are not miniature ecosystems but fragments of an ecosystem designed to simulate selected characteristics.' Microcosms, especially those in which artificial food chains are established, differ from real systems in that the experimenter selects its structure. As the response and the variability in response of a system is sensitive to its structure, the experimenter in this sense controls the response to perturbations (Crow and Taub, 1979).

Giddings (1986) supported the use of microcosms or field enclosures for defining safe levels of pollutant exposure for aquatic ecosystems. While microcosms and mesocosms may come much closer to predicting ecosystem level effects, there are still problems of scale and of missing components that Neuhold (1989) concluded would limit their use as a predictive tool. Perhaps one of their stronger critics, Moriarty (1983) concluded that microcosms 'do appear to have the worst of both worlds: they are too complicated to give results that are easily interpreted, but too artificial to be of immediate relevance to field situations.'

Another of the problems often cited in relation to microcosm studies is a low signal-to-noise ratio in system response, which impedes the detection of a significant treatment effect. Crow and Taub (1979) suggested several strategies for improving detectability in microcosm studies. Increasing both the sample size and the number of replicates will improve the power of statistical tests (Chapter 4). Decreasing the number of treatments, that is, having more replicate microcosms at fewer concentrations, appears to result in minor improvements in detectability. The authors demonstrated that, for a fixed number of microcosms, designing the experiment as a two-way analysis of variance can provide increased power relative to one-way factorial designs.

Crow and Taub (1979) discussed a number of methods for reducing the variability among replicate microcosms. Of those discussed, selection of the 'right' measurements appeared to be the most effective strategy for improving replicability among samples. The authors argued that functional parameters of microcosms are less variable and more responsive to treatments, in other words, more statistically powerful, than measures of community structure. In contrast, they found the number of individuals per species to be highly variable and tests based on this parameter therefore to be statistically less powerful. These conclusions differ markedly from those of Gray (1989) who, in agreement with Schindler (1987), argued that measures of function would provide a poor measure of impending ecosystem damage because of feedback mechanisms by which function is conserved in the face of perturbations. It cannot be overemphasized that, given the inherent problems with variability and replicability in microcosms and field observation studies, greater attention must be given to the inclusion of adequate controls and replicates in these test systems.

It is generally perceived that the concentration of a pollutant causing a change in ecosystem structure or function approximates that concentration affecting its most sensitive members. Davies and Gamble (1979) measured the effect of mercury addition in marine field enclosures. They found that the only observable effect of the addition of 1 μg l^{-1} was a temporary reduction in photosynthetic carbon uptake per unit chlorophyll. A further addition of 10 μg l^{-1} caused substantial reductions in the number of zooplankton and changes in community structure of zooplankton and phytoplankton. These authors concluded that contrary to the idea that ecosystems are as sensitive as their most sensitive members, the structure and function of intact ecosystems may actually be more robust to low levels of pollution than would be indicated by analysis of single species in isolation. A number of other studies have found microcosms to be less sensitive to pollutants than expected on the basis of toxicity tests of single component species (e.g., Pratt *et al.*, 1989), however higher sensitivities of microcosm response to pollutants have often been documented as well (reviewed in Crane, 1990). The reasons for such differences are not obvious, and may result, at least in part, because the basis for comparison of results from the two types of test system is inappropriate (Niederlehner *et al.*, 1986).

Lampert *et al.* (1989) compared the effects of the herbicide, atrazine, on plankton systems, tested at different levels of complexity. Literature data on the toxicity of atrazine to *Daphnia* indicated an LC_{50} of approximately 10 mg l^{-1} and a 50% inhibition in feeding occurring between 1 and 3 mg l^{-1}. Significant reductions in growth and reproduction occurred at approximately 2 mg l^{-1} atrazine. Significant reductions in daphnid biomass in an artificial food chain system were detected at atrazine concentrations an order of magnitude lower, between 0.05 and 0.1 mg l^{-1}, and were attributed primarily to effects on algae that provided the daphnid food source. Enclosure experiments with natural communities revealed by far the most sensitive measures of atrazine impact. Community responses were detected at atrazine concentrations as low as 0.1 μg l^{-1}. Community effects were similar at all concentrations between 1 and 100 μg l^{-1} atrazine. Primary production decreased following a lag period; as algae died, chlorophyll *a* decreased and bacterial growth was enhanced; a decrease in oxygen and food availability led to changes in zooplankton composition and abundance. At exposures of 1–100 μg l^{-1}, 90% or more of the atrazine added to the systems was still present after 18 days. At 0.1 μg l^{-1}, measurable concentrations of atrazine were not detectable after 10 days. Although the algal community appeared to recover after about two weeks, *Daphnia* were nearly eliminated in the atrazine-treated enclosures. Thus, although *Daphnia* are not highly susceptible to direct toxicity by atrazine, they can be influenced by concentrations four to five orders of magnitude lower than their LC_{50} as a result of disruptions in their food supply. As Lampert *et al.* (1989) recognized, natural populations of zooplankton are often food limited, and even a modest reduction in their food supply may push the energy balance of zooplankton populations below that necessary for maintenance. Their results clearly demonstrate that routine single-species tests can seriously underestimate the ecological impact of chemicals on natural systems.

Cairns and Mount (1990) have commented that increasing pressure from animal rights groups to reduce or eliminate the use of animals in toxicity tests may lead to increasing use of microcosms using microbial organisms. These systems have the advantage of greater complexity than standard test systems and are more amenable to the simultaneous study of the fate and effects of chemicals in biotic and abiotic components of the system. There is some evidence that microorganisms respond to similar kinds and amounts of chemicals as common test species, and that the results of microcosm tests may be easier to extrapolate to natural field conditions (Cairns and Mount, 1990). Microcosms designed to include trophic levels higher than the microbial can provide even greater information of the simultaneous fate and effects of chemicals in simplified ecosystems, albeit usually at a greater cost and on a longer time scale.

3.2.5 Community function

It has been argued that, due to feedback, functional attributes of ecosystems are poor indicators of early pollutant effects (Schindler, 1987; Gray, 1989). Studies that find no change in the rate of community respiration, primary production or other function before and after exposure to a toxicant are generally interpreted to mean that the toxicant in question has no effect on community function. However, it is important to distinguish the reasons for which a lack of functional change may occur. On the one hand, there is the case in which physiological, biochemical, behavioural and other processes within and among organisms in the community are in fact unaltered by the toxicant, and no functional change is inferred. On the other hand, significant effects of toxicant exposure may result in physiological acclimation or genetic adaptation such that overall community function is conserved. (Changes in the number or kind of species in the community, that is, structural changes, do not necessarily occur at this stage.) The exposed community is functionally different to the extent that it has become more tolerant to toxicant exposure, possibly at some cost. If we were to compare the function of this exposed community with that of an unexposed, and presumably non-tolerant community, measured at the same pollutant concentration, we would expect to observe a difference in function reflecting the pattern of tolerance. It is likely that the nature of the observed functional change is dependent on exposure concentration and on time since exposure, relative to the generation time of the populations making up the community. There is some evidence that this is true for phytoplankton communities (H. Blanck, personal communication). Upon initial exposure, rates of community photosynthesis decrease. During later stages of exposure, photosynthetic rate may be restored to approximately pre-exposure rates as the community evolves functional tolerance to the pollutant.

The reason that it is significant to distinguish between the different causes of functional consistency is that a lack of change in community function can be indicative of either 'no pollutant effect' or of 'a significant pollutant effect'. While a lack of change in functional rate processes, such as photosynthesis, may not be sufficient as the single criterion upon which to base assessments of pollutant effects on communities, this parameter can provide valuable information when used in combination with other measures of toxicant effect.

3.2.6 Community tolerance

Whenever a community is exposed to a pollutant, a number of changes may take place. As discussed in Chapter 1, responses can occur because individuals acclimate to pollutant-caused changes, selection can occur favouring resistant genotypes within a population, and selection among species can result in changes in the community species composition. All such changes may be evidenced as an increased tolerance of the community to the pollutant in question. Blanck *et al.* (1988) developed this idea for use on ecotoxicological studies of community-level pollutant effects. These authors proposed that **pollution-induced community tolerance** (PICT) provides 'direct evidence that a community is [or has been] affected by toxicants present in an ecosystem, and that the agents actually affecting the biota can be identified because PICT will be observed only for those compounds that exert selection pressure on the community.' PICT can be measured in laboratory concentration series or along pollution gradients in the field. Correlation between the degree of tolerance and pollutant concentration is postulated as evidence that the pollutant has resulted in selection on the community and has thus had an impact.

There are many advantages in the PICT approach.

1. Any changes within or among populations within the community are accounted for.
2. It is a general approach which does not rely on the use of any particular test species or type of ecosystem.
3. It should be more sensitive than single-species tolerance tests because tolerance begins to increase with the exclusion of the most sensitive members of the community.

One of the limitations of single-species tests is that competitive or other interactions with different species in the community are neglected. It has been shown that such indirect effects can have important consequences for the impact of a pollutant on populations and communities. An advantage of PICT is that it detects pollutant impacts resulting from indirect as well as direct effects operating within the community. PICT also distinguishes changes occurring within the community (due to direct and indirect pollutant effects) from secondary pollutant effects due to changes occurring outside the community. For example, if the effect of an insecticide on an aquatic system was primarily a reduction in the number of grazers, a test for PICT on the algal community would show no effect, even if algal processes were affected by the reduced grazing pressure. PICT can be used to calculate the **no-effect concentration** (NOEC) for the community as a whole. The NOEC would be defined as the highest pollutant concentration at which increased tolerance was not detected.

A disadvantage concerns the fact that functional changes which are related to life-history parameters (Gray, 1979) rather than physiological characters will not be detected by the PICT approach. For example, pollutant exposure can result in the predominance of opportunistic colonizers. Such species may reproduce at great rates and reach relatively high densities, but may not be particularly tolerant of the pollutant.

Increased tolerance arising from co-tolerance (that is, tolerance to compounds to which the community has not been exposed), or related to causes other than toxicant stress, would invalidate conclusions drawn from PICT. Although co-tolerance can potentially confound identification of tolerance-inducing chemicals, co-tolerance patterns may be used to investigate the mechanistic basis of toxicity. This will be true when the basis of co-tolerance is a common mode of action of two or more toxicants rather than a general defence strategy effective against a number of unrelated compounds.

Because any sample of a community includes a wider range of physiologies, morphologies, genotypes and life histories than a single population sample, it is likely (although not certain) that the variance in tolerance within a community will be much greater than within any single population of which it is composed. Variability in tolerance within communities not exposed to the pollutant in question, that is the control communities, will determine the ease with which PICT can be detected.

Blanck *et al.* (1988) concluded that co-tolerance occurs only among compounds that are closely related chemically or in their mode of action. However studies of pesticide resistance in target species suggest that this may not always be true and that co-tolerance may be a relatively common phenomenon, at least within single species or populations. Genetic studies of pesticide resistance in insect species have shown that genes conferring resistance may occur in high frequencies in populations never exposed to pesticides (Wood and Bishop, 1981). Several Australian mammals have been found to be resistant to a fluoroacetate poison used for rabbit control. Apparently the plants on which these species feed contain natural fluoroacetate, and it is believed that their resistance arose via natural selection before pesticide exposure.

In contrast, pleiotropic effects of selected genes have also been observed that decrease tolerance to other substances. For example, the oat variety, Victoria, possesses a gene for resistance to the crown rust, *Puccinia coronata*, which led to the widespread use of this variety in Iowa in the 1940s. Unfortunately, this gene possessed a pleiotropic effect which made the variety particularly susceptible to another fungus, *Helminthosporium victoriae*, which devastated the crop (Barrett, 1981).

Thus, co-tolerance may occur for chemically unrelated substances, and likewise co-tolerance among chemically related forms may not extend to all members of the chemical group. Wood (1981) explained that this occurs because the basis of co-tolerance lies not so much in the chemical relatedness of the substances as in the mechanisms that have evolved to resist them.

There is evidence that cytochrome P-450 systems, which are present in many organisms from bacteria to plants and animals, can detoxify a wide variety of substances including polycyclic hydrocarbons, halogenated hydrocarbons, insecticides, drugs, ethanol and biogenic amines (Bishop and Cook, 1981). Greater understanding of the functional basis or bases of tolerance and of the importance of general purpose resistance mechanisms is needed to determine the prevalence of co-tolerance. It is likely that interpretation problems related to co-tolerance will be more important for studies of single populations than of whole communities because the chance that different species in a community will exhibit the same patterns of co-tolerance should decrease as the number of species included in the analysis increases. Thus, as Blanck *et al.* (1988) suggested, co-tolerance should rarely lead to interpretation problems in community level studies.

3.3 Populations: studies of whole organisms

3.3.1 Population tolerance: distribution

Moriarty (1983) argued that although we are ultimately interested in determining the effects of pollutants on ecosystems, study at the level of the population should be emphasized. To date, ecotoxicology has focused primarily on population level effects as evidenced by the almost exclusive emphasis on the study of population tolerance distributions (as measured in LC_{50} and EC_{50} tests).

Tolerance is used here to mean an organism's ability to function successfully during exposure to an environmental stress, such as a toxicant. Following Weis and Weis (1989) among others, we use the terms tolerance and resistance interchangeably. In practice, short-term tolerance is usually equated with survival, whereas long-term tolerance can be indicated by continued growth and reproduction. Because individuals in a population vary in tolerance due to genetic and environmental influences, the tolerance of any population is described by a tolerance frequency distribution – the tolerance curve (Figure 3.1a). The median of the population tolerance curve, the LC_{50} (median

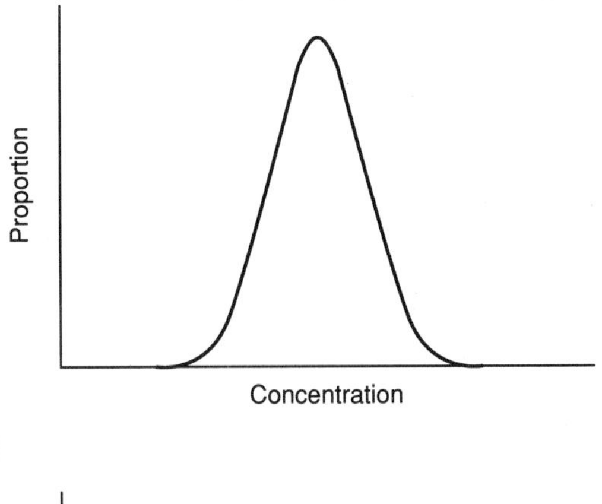

(a)

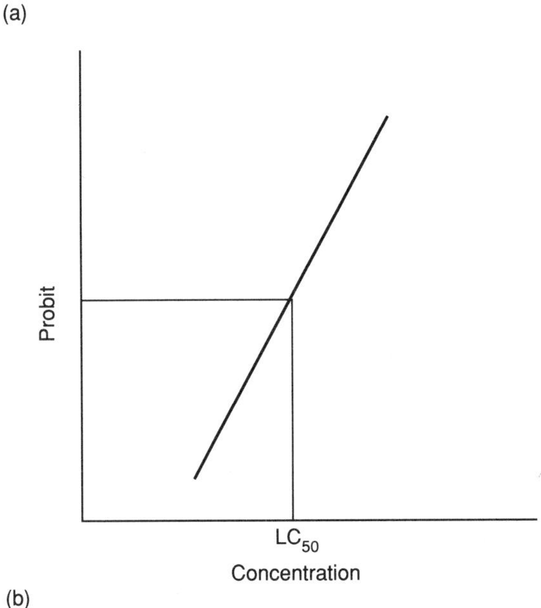

(b)

Figure 3.1 Concentration–response curves. (a) The population tolerance curve. Abcissa is chemical concentration and ordinate is the fraction of the population affected at each concentration. (b) The probit curve. Linear transformation of the population tolerance curve. The LC_{50} (or EC_{50}) is the chemical concentration at which 50% of the population is affected.

lethal concentration) or EC_{50} (median effect concentration) represents the most widely used index of population tolerance (Figure 3.1b). Virtually any sublethal effect can be used to estimate a population EC_{50}. In LC_{50} or concentration-mortality tests, death is used as a measure of effect. LC_{50} tests have been heavily ridiculed for their lack of subtlety and ecological relevance. Others strongly defend their use (e.g., Alexander and Quick, 1985). 'Although this criterion possesses recognizable limitations, it provides the first and most statistically reliable measure of the toxic effect a chemical may impose on an organism' (American Institute of Biological Sciences, 1978).

Because LC_{50} tests have reached such widespread use in ecotoxicology and because they are so strongly heralded by some while strongly criticized by others, we devote most of Chapter 4 to a thorough treatment of their usefulness, limitations and statistical analysis.

3.3.2 Population tolerance: evolution

Whereas toxicological studies have primarily used the median of the population tolerance distribution as a bioassay of chemical activity, agricultural studies have considered changes in the shape and location of population tolerance distributions as a means of studying the evolution of pesticide resistance in pest species. Unfortunately, relatively few ecotoxicological studies have made use of the population tolerance distribution to investigate the evolution of resistance to chemical toxicants, other than pesticides, in natural populations.

Klerks (1987) reviewed the literature on resistance to metals in aquatic organisms. He found that many studies reporting differences in tolerance did not distinguish between physiological acclimation and genetic adaptation. In general, studies of bacteria, algae and fungi demonstrated a genetic basis to resistance, whereas for studies of metazoa, a distinction between acclimation and adaptation was rarely made. Most assemblages occupying metal-polluted habitats showed increased resistance to the metals, resulting from some combination of physiological acclimation, changes in species composition and genetic adaptation. However, Klerks (1987) noted that the incidence of tolerance may be overestimated by a bias against publishing negative results. Incidents in which tolerance does not occur are less likely to be reported. Also, the frequent reduction in species diversity in polluted habitats would argue that many species do not become resistant to pollutants. Bradshaw (1984) showed that some species do not possess the appropriate population genetic variation for tolerance to certain stresses, and these species are eliminated from toxicant-contaminated sites.

Klerks and Levinton (1989) investigated the evolution of pollutant tolerance in a number of benthic estuarine invertebrates. Their study site, Foundry Cove, New York, USA, is probably the most cadmium-polluted aquatic habitat known and has been polluted for about 30 years. Measurements of total macrofaunal abundance showed no difference between Foundry Cove (at a number of sites of varying sediment cadmium concentration) and an unpolluted control site (Figure 3.2a). However, the number of taxa was significantly lower at the most heavily polluted of the Foundry Cove sites (Figure 3.2b). Sediment cadmium concentrations varied from 20μg g^{-1} dry weight sediment at the control site to 500 μg g^{-1} (Foundry 1), 7000 μg g^{-1} (Foundry 2) and 52 000 μg g^{-1} (Foundry 3) at the Foundry Cove sites. Laboratory toxicity experiments using the oligochaete, *Limnodrilus hoffmeisteri* (the most abundant species at the cadmium-contaminated sites),

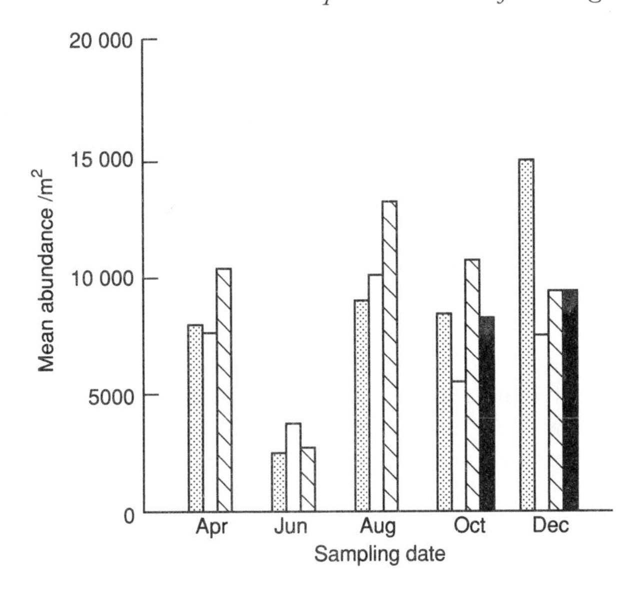

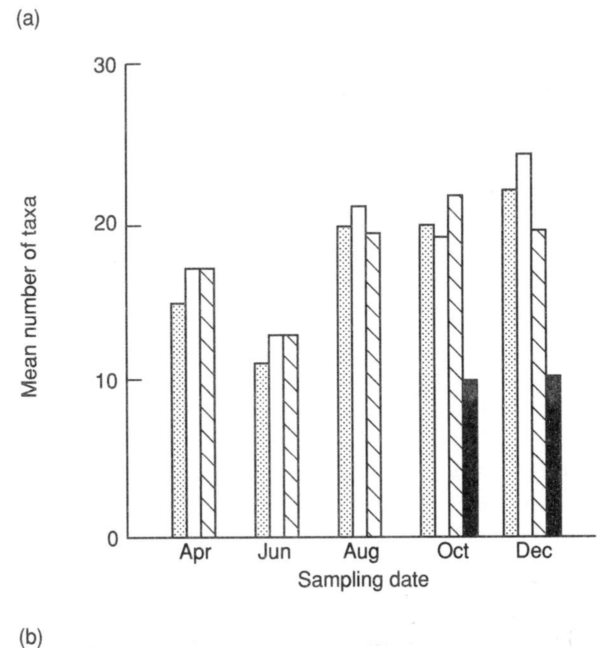

Figure 3.2 Effect of cadmium pollution on community structure of a freshwater benthic community. Sediment cadmium concentrations range from 20 μg g^{-1} dry weight sediment at the control site (▩) to 500 μg g^{-1} (foundry 1, ☐), 7000 μg g^{-1} (foundry 2, ▧) and 52 000 μg g^{-1} (foundry 3, ■) at the Foundry Cove sites. (a) Total macrofaunal abundance and (b) number of taxa at field sites and sampling dates. Redrawn from Klerks (1987).

determined that the worms from Foundry Cove had an increased tolerance to cadmium, that the increased tolerance was not due to reduced uptake of the metal (Foundry Cove worms actually took up more cadmium than control worms in laboratory experiments), and that the increased tolerance had a genetic basis. In contrast, the chironomid, *Tanypus neopunctipennis*, which was also present in high densities at the most polluted site, showed no increase in resistance to cadmium (Klerks, 1987).

3.3.3 The importance of life history

Many factors complicate our ability to identify and predict a population's response to pollutants. If stresses occur repeatedly, a population's response will depend on whether it has been able to completely recover from previous impacts. Furthermore, a population's response to pollutant impact may be very dependent on the timing of the exposure with regard to the physiological condition and age structure of the population at the time of impact (Underwood, 1989). Although we are well aware that species and populations can differ widely in pollutant tolerance, very few general principles have been proposed to guide the prediction of tolerance differences *a priori.* To the extent that a population's response to pollutants can be predicted, it may be possible to identify particularly sensitive populations or species before pollutant exposure occurs.

Stephan (1986) pointed out that for some populations a 10 % increase in mortality of juveniles may have little effect on population size whereas a 10 % increase in mortality of mature females can have a major impact on succeeding population size. As the timing of reproduction is often tightly coupled to environmental variables, particularly in temperate species, biotic or abiotic factors that act to delay reproduction can reduce the population's reproductive success or can even prevent the population from reproducing at all (if it misses its reproductive 'window'). For semelparous species, this could have especially drastic consequences, as such organisms only have one chance at reproduction. In a recent study of the population dynamics of three populations of a semelparous, intertidal gastropod, Forbes and Lopez (1990) demonstrated that a delay in recruitment (possibly due to physical stresses in the environment) occurred in one population occupying a relatively exposed habitat, and had a substantial impact on body size distribution and subsequent recruitment of this population. This caused an alteration of the growth pattern, time to reproductive size and recruitment timing compared to other local populations of this species. Unlike snails from more protected habitats, which reached reproductive maturity in the autumn following recruitment, snails from the exposed site did not mature until the following spring. This acted to delay recruitment in this population by about 1.5 months. Whereas iteroparous species can pass up a chance at reproducing if conditions are unfavourable, devote energy to growth, and reproduce at the next opportunity, semelparous species (particularly if they are annuals) may not have this flexibility.

To the extent that different species vary in their life histories, the same pollutant exposure can have vastly different consequences. For example, species with widely dispersing offspring may be better able to recover from pollutant-induced reductions in population size. Also, factors such as resource requirements and behavioural abilities can have a substantial influence on species (or life stage) response to a pollutant stress. Species at higher trophic levels are often exposed to relatively high pollutant levels due to bioaccumulation in their food, and mobile species are often able to avoid or minimize exposure by leaving the affected area.

Often species are chosen for monitoring or other pollutant studies because they are abundant and widespread. Such characteristics suggest that these are not a random sample of species and therefore may show a different response to pollutant stress relative to most species (Underwood and Peterson, 1988). Also, it has been suggested that species dominating in heavily polluted habitats may be able to do so, not because of greater toxicant tolerance, but because their life histories are such that they are favoured under such disturbed conditions (Gray, 1979).

3.3.4 Behavioural measures

If sensitivity is a desired criterion, behavioural measures should provide even more sensitive indications of stress than physiological measures, as organisms tend to respond physiologically to stress only after behavioural responses fail (Slobodkin, 1968). In fact, it has been suggested that behavioural changes are among the most sensitive response patterns yet available (Kinne, 1980). Studies of the chemosensory detection of novel organic compounds by marine species found very high sensitivities (Kittredge, 1980). Kittredge suggested that the capability of marine species to detect novel organic compounds evolved as a response to the use of allelochemical defence mechanisms by many marine species.

It is important to determine the difference between the concentration of pollutant that is detectable by an organism and the concentration at which damage commences. It is not certain that the lowest concentration which an organism can detect is also the lowest concentration causing harm. In contrast, detection may occur for some organisms or for some chemicals at concentrations well below those that would cause damage. For a behavioural change to occur in response to a pollutant, three requirements must be fulfilled (Olla *et al.*, 1980). First, the organism must be able to sense the pollutant. Second, it must be able to recognize the pollutant as harmful. Finally, the organism must have the ability or machinery to respond. If any of these steps are missing, the result will be either no response or an inappropriate response. Olla *et al.* (1980) argued that although behaviour is generally assumed to be adaptive, it may not be with regard to novel stimuli. They summarized a number of cases in which organisms were not able to sense, or were even sometimes positively attracted to, novel substances that were clearly toxic.

Kittredge (1980) classified four types of behavioural response: detective, appetitive, defensive and aberrant. Detective behaviours are difficult to determine in many organisms, but when present, indicate that the organism can sense a change in its environment. According to Kittredge's classification, appetitive behaviour would include feeding activities, mating and other adaptive behaviours. Defensive behaviours are defined as adaptive responses that allow organisms to tolerate temporary stresses imposed on them by the environment. Finally, aberrant behaviours indicate injury to sensory or other organ systems, and will persist even after the stress is removed. Thus, assays in which aberrant behaviours are observed are insensitive in that such tests indicate levels of stress that are beyond the tolerance of the organism. Although the difference between, for example, defensive behaviour and aberrant behaviour is often distinguishable, it is not always so. Complicating this further is the fact that the ecological implications of a given behaviour can vary depending on the time frame of the pollution exposure. For example, the response of the tautog, *Tautoga onitis*, to thermal stress is a reduction in activity and an increase in shelter-seeking behaviour. Whereas this behaviour is highly adaptive for short periods, it becomes maladaptive if the thermal stress continues for a long time (Olla *et al.*, 1980).

3.4 Populations: physiological and biochemical studies

3.4.1 Advantages and limitations

Many biochemical and physiological bioassays have been developed that yield more sensitive indicators of damage than death. Recently, these suborganismal measures of pollutant effect have come to be known as **biomarkers**. In the following paragraphs, we discuss some of the most common biomarkers and direct interested readers to McCarthy and Shugart (1990) and Peakall (1992) for a more complete treatment of this topic.

As the level of stress increases, organism response shifts from a normal phase, to a reversible phase, and finally to an irreversible phase which precedes death. According to Lee *et al.* (1980) biochemical approaches for assessing stress attempt to identify the reversible response range in order to detect the effects of stress before permanent physiological damage ensues.

Stegeman (1980) noted that to make a valid inference about environmental condition on the basis of some biochemical character, one must have knowledge of the normal limits of that character, and how normal variables might influence it. Rarely do we know enough about how and why physiological and biochemical processes vary in organisms in nature. Nevertheless, such measures can provide an important component of field assessment of pollutant impact. They may be especially useful when we do not know the exact chemicals to which organisms in nature are exposed. We know that interactions between pollutants and environmental variables or between different pollutants can have a profound influence on ecosystem impact. The complexity of natural systems limits our ability to appropriately test for and correct interaction effects. In such situations, the most effective and economical approach may be to evaluate effects as a whole.

Unfortunately, many physiological and biochemical bioassays proposed for use in environmental impact assessment face a number of limitations. These include specificity of response to a species or phyletic group, differences between species, lack of consistent response, slow response time, lack of ecological relevance, seasonal variation and low levels of precision. Thus, these measures may require prohibitively large sample sizes, expensive or complex equipment and may be unsuitable for application in the field (Ivanovici, 1980b). Lee *et al.* (1980) and Uthe *et al.* (1980) added to the list of limitations inherent in many physiological and biochemical techniques used in ecotoxicology. Biochemical analyses are generally restricted to organs or body fluids that can be easily sampled in sufficient quantities. Such analyses can be difficult to interpret since, for example, there is no reason to expect a relationship between the biochemical response of a specific organ and the LC_{50} (Uthe *et al.*, 1980). Although it may be relatively straightforward to correlate the concentration of chemical in an organ with a pathological condition, it may be much more difficult to determine whether the concentration we measure is caused by the pathological condition or causes it. Uthe *et al.* (1980) also argued that determining whether the observed effect is in fact pathological requires that the response be irreversible, that is, lies outside the normal stress response range. This is in contrast to Lee *et al.* (1980, see above) and would limit the sensitivity of such measures.

Physiological and biochemical tests are generally restricted to those few species that can be maintained in the laboratory and that can withstand a substantial amount of handling. Thus, it is not likely that identification of those species most sensitive to pollutants can be achieved using physiological or biochemical techniques. If physiological or biochemical techniques are used as part of a field monitoring programme, it is often recommended that they be used in conjunction with other (histological, pathological, behavioural) methods for assessing organism health.

3.4.2 Biochemical measures

Analysis of cellular response to physical or chemical stressors has been performed using cytochemical methods, which combine observations of structural and functional alterations in response to pollutant exposure. Two cytochemical stress markers, proposed by Moore (1980) are measurements of lysosomal stability and NADPH neotetrazolium reductase activity. Lysosomes are known to sequester a variety of toxic substances and to weaken or break apart in response to stress. NADPH neotetrazolium reductase is believed to be linked with mixed-function oxygenases (MFOs), which are known to metabolize many organic xenobiotics including pesticides and aromatic hydrocarbons.

The heme proteins catalysing mixed-function oxygenase reactions, namely cytochromes P-450, aid in the metabolism and excretion of toxic substances by altering the substance through the addition of one or more polar groups. The cytochrome P-450-mediated mono-oxygenases are enzymes known to metabolize a wide variety of lipophilic compounds, including steroids, fatty acids, prostaglandins, vitamins, polycyclic aromatic carcinogens, organochlorine insecticides and polychlorinated and polybrominated biphenyls (Walker, 1980). Metabolism of foreign compounds by MFOs converts toxic hydrophobic molecules into less hydrophobic chemical species that are more available for excretion. However, MFOs can facilitate the formation of mutagenic intermediates with the potential for damage to the genome (Stegeman, 1980). Systems involving cytochrome P-450 are induced by and can cope with toxicants up to a certain threshold, above which lethal effects appear. Presumably there is variability in efficiency among P-450 molecules. If so, it would be expected that natural selection could alter the toxicity threshold (Bishop and Cook, 1981).

Monitoring for mixed function oxygenase activity has been proposed, particularly for the detection of effects of organic contaminants (Stegeman, 1980). Fossi *et al.* (1988) examined the role of the MFO system in the ability of black-headed gulls to tolerate a polluted environment. They found MFO activities several times higher in gulls feeding in a rubbish dump than in those feeding in a relatively unpolluted lagoon. MFO activities were positively correlated with levels of organochlorine residues found in bird tissues. Although MFO activity was shown to be induced by exposure to xenobiotics, it is possible that genetic differences among gull populations contributed to differences in MFO activity among populations (Fossi *et al.*, 1991).

According to Ivanovici (1980b), adenylate energy charge (AEC), which reflects the metabolic potential available to an organism, may overcome some of the difficulties inherent in many physiological and biochemical approaches. It is precise, it shows a rapid response time (within 24 hours), and has been found to respond well before certain behavioural or physiological changes. The degree of variability among individuals is comparable in field and laboratory samples (Ivanovici 1980a). The reduction in AEC in response to stress appears to be general for a number of species, it appears to be independent of seasonal, tidal and spatial differences, it is amenable to both field and laboratory measurement, and early results have suggested that reduced AECs correlate with stressful field conditions. AEC has been correlated with growth, reproduction and viability (Ivanovici, 1980a). The limitations of this measure include laborious and time consuming methodology, and limitations with regard to species selection and interspecies comparisons. Other limitations include a lack of information regarding the relationship between AEC and various components of fitness (such as adult fecundity and offspring viability) and the inability of AEC to provide more than a general indicator of stress.

A great deal of attention has been directed toward study of the mechanisms of detoxification and storage of metal pollutants. Storage granules, detoxification cells such as epithelial cells, and metal-binding proteins have been shown to play important roles in regulating metal handling and toxicity. Metallothioneins are nonenzymic low molecular weight proteins that are rich in sulphur and metal (especially cadmium, zinc and copper). They are believed to play an important role in metal metabolism and detoxification (Goyer, 1991). Simkiss and Mason (1983) recommended that the term metallothionein be reserved for inducible cytoplasmic proteins with a low molecular weight (6000–10 000 daltons), a high content of cysteine (up to 30%) and a strong binding affinity for metal ions. These proteins have been found in vertebrates, invertebrates, microorganisms and plants. Since metallothioneins are known to be induced by exposure to metals, their presence in high levels may be indicative of metal exposure. However, metallothionein concentration has been found to be widely variable among species and differs among tissues and as a function of age. Although variability in metallothionein concentration may not be an ideal biomarker of metal pollution, the induction of metallothionein has been associated with increased metal tolerance (Roesijadi *et al.*, 1982; Roesijadi and Fellingham, 1987; Benson and Birge, 1985). Thus variability in metallothionein synthesis capacity, among genotypes, populations or species, might be used as a tool for predicting the relative susceptibility of organisms to metal pollution.

Other biochemical characters have received less widespread use as pollutant indicators. For example, the monitoring of changes in hormone metabolism has been suggested as a sensitive method for indicating the effects of sublethal levels of pollutants in vertebrates (reviewed in Peakall, 1992). Several stress proteins (in addition to metallothioneins) also show great promise as general stress indicators (Sanders, 1990).

3.4.3 Physiological measures

Measures of whole organism physiological function, such as respiration and ion regulation, have been used for the assessment of sublethal pollutant effects. Changes in respiration are often estimated as oxygen consumption (e.g., Atkinson, 1973; Bayne, 1975; Hutcheson *et al.*, 1985), but can also be measured as metabolic heat production (e.g., Pamatmat, 1978; Gnaiger, 1983; Schumway *et al.*, 1983) or release of radiolabelled carbon (e.g., Famme and Kofoed, 1982; Cammen, 1985; Forbes and Lopez, 1989; Forbes and Depledge, 1992b). Exposure to chemical pollutants may increase (e.g., McKenney and Matthews, 1990) or decrease (e.g., Forbes and Depledge, 1992b) aerobic respiration rates. Ion regulation has been shown to be disrupted by a variety of metals (e.g., Bjerregaard and Vislie, 1986) and organic pollutants (e.g., McKenney and Hamaker, 1984).

Bayne (1980) emphasized that, for changes in the rate of physiological functions to provide useful measures of pollutant effect, it is necessary to determine that the rate change is associated with damage to the individual or population. By this definition, damage is associated with a reduction in fitness. Thus, any physiological effect of a pollutant is considered to be significant if it imposes a stress on an organism that results in a reduction in fitness. Such a definition implies, first, that we can measure fitness and, second, that we can directly link biochemical or physiological changes with changes in fitness. Also, Bayne's approach assumes that changes in individual fitness are closely coupled with population recruitment. Underwood and Peterson (1988) noted that for many marine invertebrate species this assumption is invalid.

Linking physiological changes with changes in fitness is particularly critical given that changes in different physiological processes may compensate for each other so as to maintain normal functioning of the integrated system. For example, a reduction in feeding rate may be compensated by an increased absorption efficiency of ingested food such that absorption rate is maintained. As Bayne (1980) remarked:

> This constraint of having to demonstrate a decline in the animals' fitness is particularly important in deciding upon indices of use in measuring the effects of pollution. There is a danger that we may be seduced into measuring physiological processes simply because they are easily measured, and then recording any change as detrimental and as evidence of pollution.

Linking changes in physiological or biochemical structure or function with changes in fitness remains as one of the more significant challenges in the continuing development of pollution biomarkers.

3.4.4 Scope for growth

Measurements of growth (or the energy available for growth) have proven very useful as integrated measures of organism performance. Growth can change as a result of alterations in behaviour, metabolism or other physiological processes. It can affect time to maturity, senescence and reproductive potential. As growth is related to birth and death rates, it can have an important influence on population structure.

Bayne *et al.* (1985), Widdows and Johnson (1988), Koehn and Bayne (1989) and others have advocated the use of physiological energy balance (the energy available for reproductive and somatic growth) to assess the degree of stress to which an organism is exposed. By measuring rates of feeding, respiration and excretion, and efficiencies of food absorption, it is possible to calculate the energy (or other 'currency' such as carbon, nitrogen etc.) available for growth (the **scope for growth** (SFG) *sensu* Warren and Davis, 1967). As growth integrates the various processes of energy intake and expenditure, it should show a response when any of its component processes are disrupted. Thus, effects of a pollutant on any process related to the uptake or use of energy should be detectable as a change in energy balance. Analysis of the component processes can yield significant insight into mechanisms of toxicity. Other advantages of this approach are that for some organisms, energy balance is easier to measure than growth. This is particularly true for slow growing organisms or those that 'grow' by packing energy into high density lipids (or gametes). It also has the advantage of not requiring repeated measurements over time. To calculate a growth rate for an individual usually requires measurements to be made at least twice (start size, end size, but see Kaufmann, 1981), unless a readily interpretable record of growth is preserved in hard parts. This assumes that organisms can be marked in some way and recovered; often this is impractical. Components of the energy budget can be measured from a single time collection, making this technique suitable for biomarker studies.

Widdows and Johnson (1988) calculated scope for growth for *Mytilus edulis* from different sites in the Langesundfjord using measurements of feeding rate, food absorption efficiency, respiration and excretion. SFG decreased in mussels exposed to a mixture of petroleum hydrocarbons and copper in laboratory mesocosm experiments. SFG did not differ between the control (3.0 $\mu g\ l^{-1}$ total aromatic hydrocarbons (THC) + 0.5 $\mu g\ l^{-1}$ Cu) and the low pollutant treatment (6.4 $\mu g\ l^{-1}$ THC + 0.8 $\mu g\ l^{-1}$ Cu), but was significantly lower in the medium (31.5 $\mu g\ l^{-1}$ THC + 5.0 $\mu g\ l^{-1}$ Cu) and high (124.5 $\mu g\ l^{-1}$ THC + 20.0 $\mu g\ l^{-1}$ Cu) pollutant treatments (Figure 3.3a). SFG also decreased along a pollution gradient in the Langesundfjord and appeared to be primarily related to the concentration of petroleum hydrocarbons and PCBs (Figure 3.3b). In both cases, the decrease in SFG was determined to be due primarily to a reduction in clearance rate.

Bayne *et al.* (1985), Widdows and Johnson (1988), Koehn and Bayne (1989) and others have stressed the usefulness of the energy balance approach for its early-warning capacity. The idea is that physiological measurements will be more sensitive indicators of stress than measurements of growth itself, and will be likely to show significant effects of stress before measured growth rates are significantly reduced. Often mentioned disadvantages include the fact that SFG is a general measure of stress response, and is thus difficult to attribute to any particular environmental effect, and that SFG can be quite seasonally variable (Bayne, 1980; Gilfillan, 1980).

The choice of currency for measuring SFG can be critical. Ideally, we would want to choose a currency that is widely required for metabolism and that is limiting. Sibly and Calow (1986) suggested that energy represents the most generally suitable currency; however, rarely is energy measured directly in estimates of SFG. For example, feeding rates (i.e. for suspension feeders) are often calculated from the rate of clearance of algal cells from suspension (e.g., Poulsen *et al.*, 1982); absorption efficiency has been measured by comparing the ratio of organic to inorganic matter between ingested food and faeces (e.g., Widdows and Johnson, 1988); respiration is generally measured as oxygen consumption and excretion as ammonia loss (e.g., Bayne and Newell, 1983).

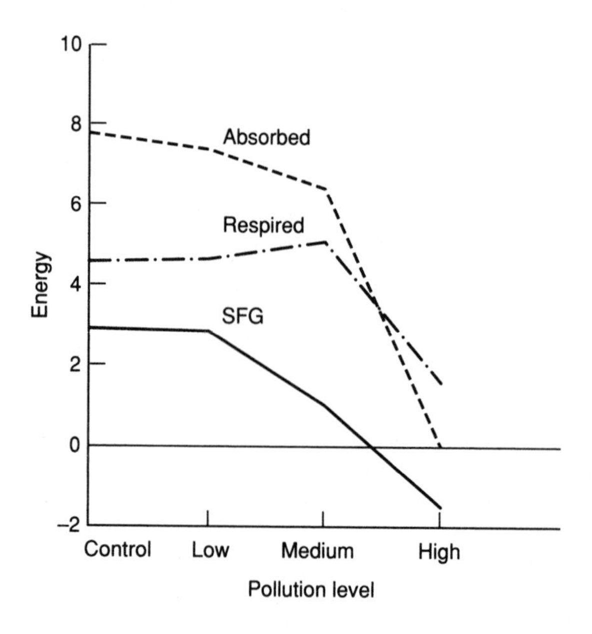

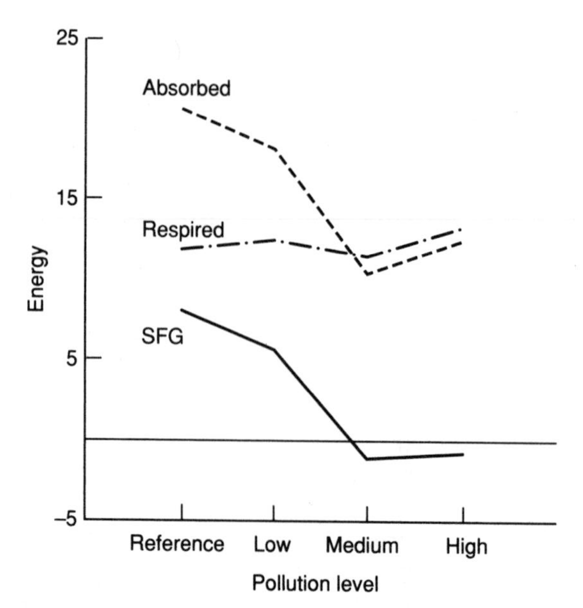

Figure 3.3 Effect of chemical pollution on scope for growth (SFG) and its components. (a) Results of a laboratory mesocosm experiment. (b) Measured SFG along a pollution gradient in Langesundfjord, Norway. Redrawn from Widdows and Johnson (1988).

Single and dual-label isotope techniques have proven quite useful in calculating carbon budgets, particularly for very small invertebrates (Calow and Fletcher, 1972; Kofoed, 1975; Lampert, 1977; Lopez and Crenshaw, 1982; Lynch *et al.*, 1986; Forbes and Lopez, 1989). More recently, direct calorimetry has been used to calculate metabolic energy expenditure (e.g., Widdows and Hawkins, 1989).

Gilfillan (1980) stated that virtually any reliable method can be used to measure ingestion and respiration rates. However, the precision and accuracy of SFG measurements can vary widely depending upon which 'reliable' methods are used. Each of the above techniques entails a series of conversion factors upon which accurate estimates of SFG rely. For example, when respiration is measured as oxygen consumption, it is often necessary to assume a respiratory quotient (the volume of carbon dioxide produced divided by the volume of oxygen consumed) for converting respiration to units of carbon. Gilfillan (1980) claimed that in comparative studies (SFG from a polluted site compared with SFG from a clean site) slightly under- or overestimating the respiratory quotient will not introduce significant error. This assumes that the respiratory quotient is, itself, not influenced by the pollutant, an assumption which is not always valid (Forbes and Depledge, 1992b). The same is true when measured chlorophyll *a* values in the food source are converted to units of carbon by assuming a constant carbon to chlorophyll *a* ratio (Gilfillan, 1980). As both of these ratios are known to show variability in response to physiological or environmental factors, it would seem that greater efforts should be devoted to determining whether such factors vary systematically in response to pollutants.

Whereas instantaneous measures of physiological processes of feeding, respiration etc. are certainly more sensitive than growth rate measurements, this does not necessarily mean that they will be better indicators of incipient stress. Very small differences in ingestion rate or oxygen consumption may not be detectable by current physiological techniques, but may nevertheless add up to considerable differences in net energy gain (growth) over time. Sensitive and short-term physiological measurements exhibit a wider degree of interindividual variability than their longer term composite (growth). While such variability is interesting in its own right (and we suggest it deserves further study), it can obscure differences among treatments. For example, Forbes and Depledge (1992b) investigated the effects of cadmium exposure on growth and carbon balance in an estuarine gastropod species. They found that growth rates were substantially reduced by exposure to 100–200 $\mu g\ l^{-1}$, such that after one month of exposure, unexposed snails were more that twice as large as exposed snails. The authors estimated that to obtain such a size difference would require a difference in daily net carbon gain of only 5 $\mu g\ day^{-1}$. Althouth such differences can be detected using sensitive radioisotope techniques, the degree of variability among individual snails, when measured at this scale, is much greater relative to the effect size, making differences among groups very difficult to detect statistically.

Predicting the consequences of pollutant stress on a population, using calculations of scope for growth, is complicated by uncertainty in the way organisms partition available energy between growth and reproduction and by our lack of knowledge as to which and in what proportions different energy needs are met by incoming food versus stored energy reserves. The former can be of particular relevance to organisms having indeterminate growth. It is conceivable that under stressed conditions some organisms could exhibit a reduced scope for growth but maintain an equivalent reproductive output if reproduction is conserved at the expense of growth. To the extent that this occurs, scope for growth will be at least temporarily decoupled from fitness. Other species seem to conserve growth (or size) at the expense of reproduction. In such organisms a reduction in scope for growth more closely approximates a reduction in reproductive output. Female *Capitella* sp. 1 respond to stress from starvation and hypoxia by resorbing their eggs. Thus, when incoming energy is close to zero, females have lower starvation rates than males (Forbes, 1989). Organisms are likely to differ in the minimum energy surplus required before they begin or continue producing gametes. Such differences may be related to, for example, age, habitat or life history. There is no reason to expect that such thresholds are not themselves subject to selection and thus may differ between polluted and unpolluted habitats.

Although scope for growth has been shown to correlate with actual growth in species such as *Mytilus edulis*, equations relating SFG to growth or fitness in a variety of species is currently lacking (Grant and Cranford, 1991). Gilfillan (1980) strongly advocated the use of SFG for monitoring the effects of pollution. He wrote, 'When periodic measurements of scope for growth are made for two or more populations of animals the resulting time series data allow prediction of the relative growth rates of each of the populations.' This statement assumes that populations from polluted and non-polluted habitats are equally efficient at turning absorbed food resources into tissue. It neglects the possibility that there are significant energetic costs to detoxification. If an organism is tolerant because it allocates available energy to detoxification mechanisms, such as detoxification enzymes, rather than growth, then (assuming absorption and respiration rates to be constant) the cost of detoxification would be reflected in reduced growth rates. In such a situation SFG would not be a good predictor of actual growth, and SFG could be uncorrelated and actual growth negatively correlated with fitness. There are few data measuring tradeoffs between growth and mortality from pollutant stress; however, an interesting theoretical treatment of this topic is given by Holloway *et al.* (1990).

Underwood and Peterson (1988) noted that measures of pollutant impact at the organismal or suborganismal levels relate more to growth and reproduction than to population abundance or contribution to food webs. These authors pointed out that a problem with measures of pollutant effects at levels below the population is that they may lack predictability because of a decoupling between individual performance (growth, reproduction and sometimes even death) and population response. However, Calow (1988) argued that

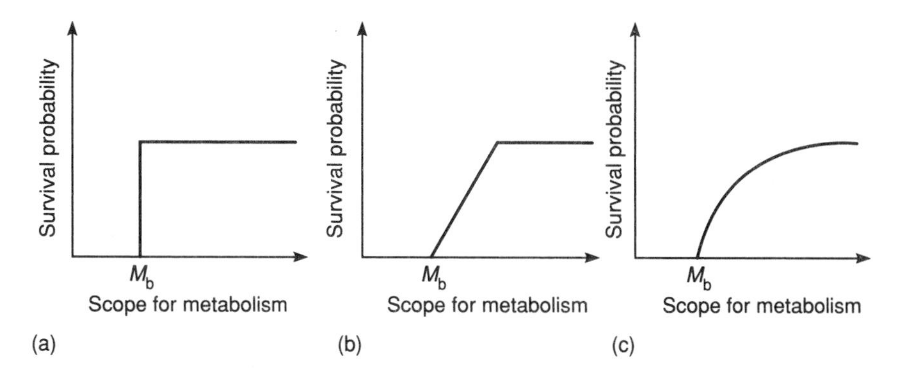

Figure 3.4 Model relationships between scope for metabolism (SFM) and survival probability. (a) Above the basal metabolic rate, M_b, survival is independent of SFM. (b) Survival increases linearly with SFM. (c) Survival increases curvilinearly with SFM. Redrawn from Calow and Sibly (1990).

physiological models (such as scope for growth) are valuable tools for linking the physiological response of individuals to population-level phenomena, as well as for elucidating mechanisms of toxicological responses. An important assumption of the physiological models is that all resistance and repair processes are energy demanding, meaning that metabolic rate should increase with increasing levels of toxicant exposure, that is, concentration or time of exposure, until irreversible effects impair metabolic function itself (Calow, 1988). Calow and Sibly (1990) developed this energetic approach by partitioning absorbed energy (A) into metabolic costs (minus costs of production; M) and production (somatic growth and reproduction) plus metabolic costs of production (P_T). Since $A = M + P_T$, this implies tradeoffs between the energy to fuel metabolic costs and the energy devoted to growth and reproduction. Scope for growth can be reduced by either a reduction in A, an increase in M, or both. Several possible functional relationships between M and survival probability (S) were proposed (Figure 3.4). In the first case, S is not enhanced with increasing M above the minimum basal rate. In the second and third cases, S increases with increasing M linearly or curvilinearly, respectively. P_T is functionally related to reproductive output (that is, time between breeding periods, number of propagules). The different relationships between S and M, plotted in Figure 3.4, can be used to model the effect of a reduction in A or an increase in M on population growth. The Calow and Sibly (1990) models link metabolic measures of stress to effects on population dynamics by predicting how the SFG of individuals will translate into patterns of organism distribution and abundance. In keeping with the above caveat of Underwood and Peterson (1988), Calow and Sibly (1990) noted that for populations under density-dependent control, such factors may override physiological effects in controlling population dynamics.

3.5 Populations: genetic studies

3.5.1 Changes in gene frequencies

Organism performance is a function of both genetic and environmental factors. Genotype determines, 'not a single phenotype, but a norm of reaction, a repertoire of phenotypes that can arise in different environments or successions of environments' (Dobzhansky, 1970). Assigning a certain level of fitness to a genotype or allele thus only has relevance with respect to a specified environment. It has been well established that genotypes which exhibit superior fitness under one set of environmental conditions can be of very low fitness under another set of conditions (Dobzhansky, 1970; Mueller *et al.*, 1991).

Some evidence for the existence of genotypes that tolerate a wide range of environmental extremes has come from the study of 'vigour tolerance' in insects. It has been found that strains developed by breeding only those individuals that survive extreme environmental conditions are often better able to tolerate many kinds of stress, including exposure to chemicals, and may be more easily selected for tolerance to chemicals. However, such 'vigour tolerance' may also decrease the chemical tolerance of the selected population (Hoskins and Gordon, 1956).

Present theoretical and empirical understanding of population genetics has relied heavily on results from protein (largely metabolic enzyme) electrophoretic analyses. Although enzyme electrophoresis is easier and less expensive than DNA sequencing techniques, variability in electrophoretic allozymes provides a biased view of genetic variation. It studies only certain classes of mutation in the coding region of the structural genes for a very small subset of proteins (Brown and Burdon, 1987). Although it is remarkably difficult to distinguish an allele that is itself adaptive from one that is only linked to another trait, such marker genes can nevertheless be quite useful for providing an indicator of organism response to environmental change (Berry, 1980). However, if the goal is to investigate whether the enzyme polymorphism is adaptive, knowing the biochemical function of an enzyme, its role in metabolism, and the performance of the different allozymes is critical (Koehn *et al.*, 1983). Evidence is continuing to accumulate that electrophoretically detectable allozyme frequencies vary in response to physical (temperature, salinity, for example), chemical (pollution), spatial (e.g., with latitude) and temporal (such as seasonal) variability among habitats (Battaglia *et al.*, 1980; reviewed by Berry, 1980; Koehn *et al.*, 1983; Nevo *et al.*, 1983). Allele frequencies can change by mutation, selection, migration and genetic drift. For such changes to provide useful indicators of environmental (pollution) perturbations, changes due to adaptation must be distinguishable from those due to migration and drift (Berry, 1980). Assessing the adaptive potential of allozyme variants requires a demonstration that structural (electrophoretically detectable) differences reflect functional differences.

Furthermore, it is often desirable to relate differences in enzyme function to differences in fitness.

It has been widely debated whether the observed variation in allozyme frequencies is the result of balancing selection or is largely due to stochastic forces (Harrison, 1977). The latter holds that enzyme variants are maintained by genetic drift and are thus selectively neutral. A major focus of population genetics has thus been to link allozyme variation to genetic variation for ecological or fitness-related parameters.

There is an increasing body of evidence to suggest that enzyme polymorphisms represent phenotypic diversity and that different alleles are associated with differences in physiological performance. It remains to be conclusively demonstrated whether the observed differences in physiological function are in fact adaptive. As Koehn *et al.* (1983) pointed out, the biochemical/physiological variations that have been linked to allozyme polymorphisms represent 'potential mechanisms of adaptation, but not necessarily adaptive mechanisms.'

Harrison (1977) argued that parallel variation in electromorph frequencies among sympatric species can only be explained by balancing selection. Using the same reasoning, Nevo *et al.* (1983) proposed the use of allelic isozymes as genetical monitors of pollution. This work provides evidence that shifts in frequencies of allelic isozymes occur in response to pollutant exposure. Nevo and colleagues combined *post-hoc* correlations of allelic isozyme frequencies and pollutant levels in natural populations with controlled laboratory experimentation to test the tolerance of different electrophoretic phenotypes. Summarizing data from a number of studies on barnacles, marine crustaceans and gastropods, they determined that of 15 loci tested, allozymic variation was predominantly nonrandom and was selected by thermal, chemical and heavy metal pollution. In particular, they concluded that the frequencies of phosphoglucomutase and phosphoglucose isomerase were sensitive to and varied adaptively with the degree of pollution. Thus they suggested, 'This unique sensitivity can and should be the basis for predicting and controlling environmental quality.'

It has been well documented that exposure to environmental toxicants may exert a stress on living systems and that the result of such stress can be a change in population genotype frequencies. Bradshaw and Hardwick (1989) wrote, 'Since stress can be defined as anything which reduces growth or performance, it follows that, if appropriate genetic variability is present, classical evolutionary changes in populations are to be expected in any situation where a consistent stress is occurring.' Their paper summarized the evidence available for the evolution of resistance by plant species to a variety of environmental stresses, including heavy metals and herbicides. The work of Nevo *et al.* (1983) adds to the evidence suggesting that chemical pollutants impose a selection pressure on living organisms and that such selection can result in changes in genotype frequency within populations and a consequent increase in population tolerance. Such genetic biomarkers may indeed provide early warning of incipient pollutant impact.

In theory, it can be shown that selection applied to a fairly complicated genetic system (10–20 genes) produces no change in the total adaptive value of a population for a long time despite the fact that gene frequencies are changing (Lewontin, 1965). Thus if the genetic basis for pollutant tolerance is polygenic, even large changes in the frequency of one or a few alleles may have very little effect on the phenotypic expression of resistance. This suggests that any relationship between allozyme frequencies and pollutant tolerance might be difficult to detect because of time lags. It also means that even if we cannot demonstrate a significant correlation between allozyme frequency change and fitness it should not automatically be assumed that the change is selectively neutral.

This returns us to the question of value judgments. It can be reasonably argued that selection of tolerant genotypes within a population is direct evidence that a pollutant is having an impact on the population. For at least some of the individuals in the population this impact is decidedly negative. However, do changes in population genetic structure, which may indicate shifts in the relative fitness of genotypes within a population, necessarily indicate a negative impact on the population as a whole? For example, if we define 'negative pollutant impact' on a population as a significant decrease in net reproductive rate (combining survival and fecundity data), conceivably we could have a situation in which substantial changes in genotype frequency occur, but in which the net reproductive rate of the population does not change. If the appropriate genetic variation for pollutant tolerance does not exist within a population, then the gene pool will show no genetic change even if the population is stressed by the pollutant (Moriarty, 1983; Bradshaw and Hardwick, 1989). Also, if the gene pool of one species is affected, this does not necessarily indicate that the numbers of all other species in the habitat lacking the appropriate genetic variability will have been reduced (because such will also be determined by absolute tolerance, life history, population structure, etc.).

Much of the available genetic evidence for pollutant tolerance comes from studies of the response of insects to pesticides and studies of terrestrial plants occupying metal-polluted habitats. Although standardized laboratory tests often implicitly assume a common mechanism for sublethal and lethal effects of a toxicant, studies of knockdown and mortality of houseflies, *Musca domestica*, from DDT suggested a different genetic basis for the sublethal and lethal responses. Although knockdown from DDT was dominant and involved a single gene or group of linked genes, mortality appeared to be under multigenic control (reviewed by Hoskins and Gordon, 1956). Such independent genetic control of lethal and sublethal resistance mechanisms may help to explain the lack of consistent relationship (within either species or chemicals) between the pollutant concentration at which sublethal effects occur and that at which mortality ensues.

A study by McNeilly (1968) of copper tolerance in *Agrostis stolonifera* found that only tolerant plants were able to grow in copper-polluted soils, whereas both tolerant and non-tolerant plants grew equally well on uncontaminated soils. However, when tolerant and non-tolerant genotypes were grown together on uncontaminated soil, the non-tolerant plants grew much faster than the tolerant plants. Moriarty (1983) wrote, 'In all known instances when a species occurs in both normal and contaminated soils, metal tolerance is genetically determined, and such genotypes do not occur commonly on uncontaminated soils.'

The development of tolerance is positively related to the strength of selection and to the availability of tolerant alleles, whereas high rates of gene flow slow down the shift to tolerant genotypes. From studies of tolerance in plants, it appears that some (perhaps many) species lack the appropriate alleles that would enable them to adapt to strong selection from pollution (Bradshaw and Hardwick, 1989). Attempting to identify those genetic factors that increase the likelihood of tolerance developing may provide an important direction for future work.

3.5.2 Genetic diversity

The effects of genetic diversity have been most notably recognized in agriculture, in which vast tracts of genetically uniform varieties are often planted. This practice has been characterized by 'boom-and-bust' cycles (Barrett, 1981) during which an initially disease resistant variety is introduced and becomes widely used by farmers. Resistance breakdown often follows as disease organisms evolve mechanisms to overcome the resistance, and agricultural use of the variety declines. It has been recognized that the use of heterogeneous plant populations slows down the rate of epidemic development, and more recent practices are incorporating more heterogeneous planting strategies (Barrett, 1981). From a different perspective, studies of insecticide resistance have found that the use of insecticides against a population having a wide range of susceptibilities will select out a range of resistant individuals. Insecticide application on a population having a narrow range of resistances will either result in most of the population surviving or in the survival of very few resistant individuals who contribute little to the next generation because they often show reduced fertility. Thus, insecticides to which pest species show little variability in response have been recommended as resulting in fewer resistance problems (reviewed in Hoskins and Gordon, 1956). These authors also noted that, although it would be expected that elimination of the more susceptible individuals from a population by exposure to a chemical would result in a more homogeneous strain, this is not always observed. In fact, comparison of LD_{50} tests and log-dose probit slopes among seven insect species and six toxicants demonstrated an increase in heterogeneity (shallower slope) with decreasing susceptibility to the toxicants. Such flattening and shifts of the line to higher dosages have been interpreted as the occurrence of 'true resistance' (as opposed to vigour tolerance, which can result in minor increases in resistance) in the population. A few studies have found an increase in slope after the population had reached a high level of resistance. An increase in slope with only a slight shift of the line to higher doses or exposure concentrations is expected when susceptible individuals are eliminated but when no mechanisms for resistance are present in the population.

Rapport (1989) noted a decline in genetic diversity in the Gulf of Bothnia, and considered this one of several symptoms of ecosystem distress. How do such declines in genetic diversity occur and what are their consequences for the population? As population numbers are reduced by pollutant impact, the number of genes in the population gene pool is reduced. Presumably, the pollutant has acted as an agent of selection by eliminating those individuals in the population that were less tolerant of the pollutant's effects. Such selection can lead to increased tolerance in the population. However, if the impact of the pollutant is so severe that only a few individuals survive, these few individuals may contain a limited and biased fraction of the original gene pool of the population. Hoffmann and Parsons (1991) suggested that under severe stress, during which individuals with major genes at the extreme of the phenotype distribution will be selected, much of the genetic variance in the base population will be unimportant with regard to the population's genetic response to the stress. However, if the ability of a population to adapt to a changing environment requires genetic variability (Williams, 1975; Bell, 1982), then reductions in genetic diversity by pollutant exposure may result in increased resistance to specific pollutants, but may reduce the population's ability to deal with natural stresses or unrelated kinds of pollution. Reducing population size to a very low number increases the chance for inbreeding (in sexually reproducing species), which can have deleterious effects on the population (Futuyma, 1979). Inbreeding depression is well known to commercial breeders. Massive declines in fertility, viability and other fitness components occur as the population becomes homozygous for deleterious recessive genes (Maynard Smith, 1989). Such 'founder effects', which occur when a population undergoes a severe bottleneck in population size, may accelerate the rate of speciation (Futuyma, 1979). This may help to explain how many opportunistic species that are common inhabitants of severely polluted environments have been found to consist of a suite of genetically distinguishable sibling species (Gray, 1989).

With reference to test systems, Hoskins and Gordon (1956) suggested that the tolerance of a species should be taken as that existing in 'the typical population existing in its natural condition rather than of inbred laboratory strains.' In natural populations, inferior individuals are eliminated by selection. Such may not be the case for inbred laboratory strains. Thus, Brown (1950, cited in Hoskins and Gordon, 1956) recognized that comparison of chemical tolerance between natural populations and inbred laboratory strains may give the false impression that the former group is resistant. Also, because in large, natural populations there is a greater chance that genes for resistance can become linked to some vital function, it is less likely that reversion (loss of resistance) will occur, relative to inbred laboratory populations.

3.6 Summary and conclusions

The effects of chemical pollutants on biotic systems can be detected at all levels of biological organization. Of the many approaches that have been developed for investigating the fates and effects of pollutants in nature, there is no single approach that has proved clearly superior. Well-recognized tradeoffs between ecological realism and replicability (that is, statistical reliability) are involved in virtually every test system. Although it would be convenient if we could devise a universal stress indicator, such a goal has proved to be impractical, if not impossible. Nevertheless, several biological stress indicators have been used with success. Specific indicators have the advantage that the stress-causing agent is relatively easy to identify, although of limited use to the extent that different indicators are required for every potential pollutant. General stress indicators can be quite valuable for detecting the presence of a stress when the source is unknown or complex, but are unsuitable for identifying causative agents in the field.

Field monitoring programmes can determine rates of release of pollutants into the environment, their subsequent behaviour and biological effects. To be most effective, monitoring programmes should be established before pollutant release occurs. Monitoring programmes are not likely to be useful for predicting damage from highly toxic or non-accumulating chemicals or for pollutants whose effects are likely to be limited in time or area.

Measurements of pollutant effects at the community level include observations of community structure and function, as well as manipulative experiments using field enclosures, microcosms or artificial food webs. Practical and theoretical problems occur with many of the commonly used diversity indices, but these remain as the most widespread measure of community structural change. Multivariate analyses of community structural change are more complicated to calculate and interpret, but may be superior for analysing changes in complex systems. Manipulative experiments, including microcosms, currently provide the most powerful approach for combining the simultaneous study of pollutant fate and effects in complex systems.

Current regulations for environmental protection in Europe and North America still require single species tests as the only basis for many regulatory decisions. Given the inherent difficulties of extrapolating from single species to multispecies effects, it is important that the development of multispecies tests for regulatory purposes receives increasing support. Constraints of time and money will probably limit standard test systems to microorganisms or small invertebrates. How well such microsystems can be used to detect or predict pollutant effects on larger organisms and communities remains to be seen.

At the population level, various biochemical and physiological indicators of pollutant effects have been proposed. These measures can be quite useful for investigating the mechanisms of pollutant impact and for detecting pollutant effects – provided that enough background information is available with regard to the 'normal' range of the indicator variable and its response to changes in non-pollutant related factors, such as season, species, detection method etc. Analysis of the shape and location of the population tolerance distribution can provide a valuable tool for the study of pollutant effects. Temporal or geographical changes in population tolerance can reveal whether populations have or are likely to evolve resistance to pollutant exposure. Because pollutants can act as effective agents of natural selection, studies of population genetic structure and changes in gene frequencies can complement studies of tolerance. In addition, the effects of pollutants on genetic diversity, particularly as it affects the ability of populations to withstand future stresses, need to be more fully investigated. In this regard, the extensive literature on pesticide resistance can provide valuable insights into our understanding of toxicant resistance.

It is critical to distinguish between methods appropriate for ecotoxicological research and those appropriate for regulatory testing. Whereas all of the methods described in this chapter have been used in research, few have been employed in regulatory decision making. In the next chapter we focus on selected methods currently in wide use by regulators. Methods designed to fulfil regulatory requirements should be relatively simple to perform and be amenable to standardization. Methods designed to investigate ecotoxicological phenomena need not be so severely restricted and can thus provide a more sophisticated level of information. These two classes of methods meet very different but important needs. The failure to recognize this point has impeded co-operation among ecotoxicologists working in different sectors (government, industry and academia), has led to dichotomous arguments over how limited funds should be allocated and has hampered our ability to deal with pollutant-related problems.

4 *Room for improvement*

The concentration of xenobiotics in unpolluted groundwater is usually very low.

From a 1992 Report to the Danish Agency of Environmental Protection evaluating a report on biodegradation from the Organization for Economic Cooperation and Development, Paris

4.1 Application of toxicity testing in ecotoxicology

Clearly, there is more to coping with existence in an increasingly polluted world than applying ever more standardized and harmonized toxicity testing protocols to each new chemical that appears on the market. Although this is an important component of a sound overall approach to chemical toxicity, it is only a small part of the story. Yet because of the pressing nature of our environmental problems we are in danger of focusing too much attention on the lesser problem of toxicity test development at the expense of any deep understanding of the behaviour of toxicants in natural systems. Despite this caveat, laboratory test systems have an important role to play in the evaluation of chemical toxicity, and it is important that they be carried out in a scientifically sound way.

The experimental analysis of the toxicity of chemicals and chemical mixtures is an area where science interacts strongly with the concerns of society for public and environmental safety. Results of toxicity tests are often used directly in decision making and government regulation through their incorporation in risk assessment protocols and for the classification and labelling of chemicals. Because toxicity tests occupy a critical position between science and policy, it is important that the methods and assumptions behind the tests be clearly understood.

All toxicity tests are based in some way on a relationship between the dose (or concentration) of a toxicant and a biological response. The differences among the various tests can be expressed as differences in endpoints and response variables. Examples include the relationship between toxicant concentration and the number of eggs or viable offspring produced by water fleas over a three-week period or the acute mortality response of the same population to a xenobiotic. One type of analysis of concentration–response relationships yields estimates of the two determining parameters of a population's response to different concentrations of a toxicant. The first parameter is an estimate of the familiar LC_{50}, LD_{50}, or EC_{50} value, which gives the toxicant concentration or dose at which 50% of the population is affected. The second is the variance of the population tolerance distribution itself. The validity of the analysis is based on the assumption that population response is Gaussian (normal) and therefore the median is coincident with the mean. The response variables can be either quantal (that is, binary, effect/no effect) or quantitative (continuous or metric). In the following pages, we address the assumptions and limitations of some current toxicity test methods in order to highlight the areas in need of improvement. We do this by focusing our attention primarily on the quantal concentration–response relationship. Despite a long history of development, the analysis and interpretation of quantal data is frequently performed poorly and misunderstood by regulators and practising experimentalists alike. Because of its central position and historical importance in ecotoxicology, we focus our analysis principally on the relationship between exposure concentration and acute mortality. Our analysis of this fundamental relationship in toxicology and ecotoxicology can be used as a model for the degree of critical appraisal required before the implementation of any test.

4.1.1 Goal of toxicity tests

Toxicity tests are performed to evaluate or predict the effects of toxicants on biological systems and to assess the relative toxicity of substances, often for regulatory purposes. Some workers perform toxicity tests for the specific purpose of predicting what biological functions will be perturbed by toxicant exposure, whereas others carry out toxicity tests to explicitly quantify the effect of a toxicant on an organism's health or viability. Still other toxicity tests are used to estimate potential hazard as part of risk assessment protocols. Clearly, while different aspects of toxicity testing are emphasized, depending on their specific purpose, certain basic features are fundamental to the design and analysis of all toxicity tests. How well currently employed toxicity tests achieve each of the above goals depends in part on the depth of knowledge underlying their development and application.

4.2 Dose–effect

The classic dose–effect relationship is a fundamental, perhaps **the** fundamental, relationship in toxicology (Amdur *et al.* 1991). It is generally measured as the dose (or nominal exposure concentration) that elicits a specified effect (often death) over a fixed time interval. The dose–effect relationship can also be considered as the time to elicit a specified effect at a fixed dose (or concentration). Whether it is more appropriate to measure toxicity as a function of concentration or time depends on the specific system and purpose of the test. For example, tests that measure the time at which a toxic effect appears are useful only for species in which the exact time of death can be clearly observed. In addition to the LC_{50}, a number of more sensitive endpoints have been used, such as the effective concentration (EC), the no observed effect concentration (NOEC or NEC), the lowest observed effect concentration (LOEC) and the maximum acceptable toxicant concentration (MATC).

By dose we refer to the quantifiable amount of a chemical or chemical mixture introduced into an organism. Typically in ecotoxicological studies an exact dose is not applied, but rather the organism is exposed to a known concentration of a toxicant added to its food or environment. Thus, the nominal concentration of a toxicant, the quantifiable amount of chemical added to the environmental medium, is generally reported instead of dose. Sometimes the concentration of toxicant in the medium is measured and related to the nominal (added) concentration. Real differences between nominal and measured concentration have no doubt contributed substantially to the observed variability in LC_{50} results. Nominal concentrations continue to be reported in the ecotoxicological literature even though these values are of little toxicological relevance. Although measured concentration may be used to provide a more accurate index of exposure, differences in bioavailability can cause the actual exposure (dose) to vary even for a single measured concentration. To obtain a solid understanding of the effect of a toxicant we need to be able to quantify bioavailability of a chemical in the environment. This is easier said than done, because for many substances we have only a partial idea of what controls bioavailability. For substances that are not rapidly metabolized or excreted, the most effective and relevant approach may be to relate toxicity to the concentration of toxicant taken up by the organism.

Effect has often been found to be a log-normal function of dose. Trevan (1929) was among the first to recognize that the theory behind the log-normal dose–effect model implied that there was no dose at which all organisms in a population would be unaffected. This is based on the inherent shape of the normal distribution, which has tails extending to infinity at both of its extremes. The log-normal tolerance model has a direct bearing on the question of toxic thresholds and on the distinction between population and individual thresholds. The sigmoid nature of the cumulative response function appears to have been misinterpreted as indicative of the existence of a population toxicity threshold (e.g., Truhaut, 1977; Hayes, 1991). This interpretation is incorrect because the normal model extends to infinity in both positive and negative directions. In contrast, the tolerance distribution can be correctly interpreted as a frequency distribution of individual thresholds for the given response measured. That is, those individuals in the population responding to a given treatment concentration have reached their threshold for this response. Theoretical considerations aside, individual toxic thresholds have been clearly indicated for a number of substances. The nutritional requirement for essential trace metals in most organisms aptly demonstrates the idea of a threshold. However, for many carcinogens it has been postulated that there is no individual threshold and that initiation of the carcinogenic process may take place even at very low exposures (Kuiper-Goodman, 1989). Kuiper-Goodman also argued that non-genotoxic chemicals have no threshold and therefore safety factors for these substances cannot be determined. Furthermore, current test procedures to determine the genotoxicity of a substance may be insufficiently powerful to accurately determine low-level thresholds even if they do exist.

We emphasize that the theoretical log-normal, dose–response relationship does not imply the existence of population thresholds. The model, in fact, implies that it would be theoretically impossible to find a toxicant concentration to which no organism in a population would be susceptible. Strictly speaking, the empirical demonstration of a threshold invalidates the model. Thus, we are forced to make a choice. Either there is no threshold and the model is a useful description of the population response, or there is a threshold and a new model is needed. No doubt the answer will depend on the toxicant tested.

Sometimes an apparent threshold may be observed but may have arisen as an artefact of the experimental design. For example, if small sample sizes or samples from a narrowly distributed population are tested, those individuals in either tail of the distribution may tend to be under-represented or missing, simply due to the nature of the sampling process. It may be argued that those few individuals in the population that are highly sensitive or insensitive to a given pollutant are of relatively little importance when it comes to setting standards for environmental protection. In this case, practical considerations may warrant that an approximate threshold be defined which could be set at the 'no effect' level for, say, 95% of the population (e.g., Wagner and Løkke, 1991 and references therein). This approach may prove practical if we have a reasonably detailed understanding of the shape of the tolerance distribution of the target population. Nevertheless, it is essential to keep in mind the difference between practical guidelines and the theoretical implications and limitations of our models.

Two points are worth emphasizing at this stage. According to the log-normal dose–response model, there is no theoretical population threshold under standard theory. Thus an observed 0% response rate is simply the 'luck of the draw'. The odds of a no-response sample occurring depend on the true, underlying response rate and the sample size (Forbes, 1993). The second, and closely related point, is that our ability to detect a 'threshold' experimentally (our ability to measure a significant LOEC), will depend critically on the statistical power of our experimental design. An illustration may serve to clarify this point. If we were to repeat an experiment over and over again, changing only the number of replicates per treatment, we would find that the LOEC tended to be inversely related to the number of replicates. As the number of experimental replicates goes up, we are able to detect lower and lower LOECs. Clearly this is an unpleasant situation. We are uncomfortable (or should be) when our standard LOEC (or NOEC or MATC) is so heavily dependent on sample size. We suggest that one way out of this predicament is through statistical power analysis. We will discuss the concept of statistical power and approaches for dealing rationally and consistently with LOEC detection and related testing problems in some detail in section 4.6. Another approach to this dilemma is to focus on quantifying the relationship between concentration and response rather than using standard inferential statistics such as analysis of variance to make hypothesis tests (Forbes, 1993). Rather than testing for treatment-dependent (and therefore arbitrary) NOECs and LOECs, one could employ regression analysis to model the dependency (Stephan and Rogers, 1985). We think this strategy makes a great deal of sense and, like the power analyses required to properly evaluate the hypothesis testing approach, focuses attention on the right areas. Both approaches necessitate consideration of the biological or regulatory importance of the adverse effects observed (Stephan and Rogers, 1985; Cohen, 1988).

4.3 Problems with laboratory toxicity tests

4.3.1 Acute toxicity tests

The LC_{50} and similar measures conventionally form a major part of the assessment of ecological risks of potential pollutants. The acute lethality test has, in fact, been acclaimed as the most useful of all available toxicity tests (Cairns *et al.*, 1978). Historically, it appears that 'the LC_{50} test was designed for the biological assay of drugs that could not be determined chemically' (Moriarty, 1983). The critical features of the bioassay have been summarized by Bliss (1957). A bioassay, according to Finney (1978), is 'an experiment for estimating the nature, constitution, or potency of a material (or of a process), by means of the reaction that follows its application to living matter. [Thus] ... the main purpose of every assay is to estimate the parameter representing the potency of a test preparation relative to a standard'. Historically bioassays were developed to standardize drugs and other materials by means of the reactions of living matter. Potency is considered a property of the test substance, not of the response, and potency is measured on a relative rather than an absolute scale (i.e. the bioassay provides only a relative estimate of the true potency of a test substance). Thus, the response of an organism to a substance is expressed relative to the organism's response to some standard preparation (insulin preparations, for example, are expressed in international units).

Both the reliability and efficiency of a bioassay are linked inseparably with its design. Ideally, the substance being assayed is identical to the standard in its biologically active principle and differs only in the extent to which it is diluted by solvents or other inactive materials. Whenever the substance assayed differs chemically from the standard, the results of the bioassay will be less accurate (Finney, 1978). Although bioassays and toxicity tests such as the LC_{50} have proved quite valuable in the standardization of pharmacological substances, one of the great unsolved problems in ecotoxicology concerns the extent to which such tests can be used to evaluate higher level responses, for example, of populations, communities and ecosystems, to pollutants.

Many laboratory toxicity tests are designed to detect acute responses, where **acute** has been defined as 'a condition involving a stimulus severe enough to rapidly induce a biologic response'. In aquatic toxicity tests, a response observed within 96 h is typically considered an acute test, death being the most common effect observed (American Institute of Biological Sciences, 1978). For some reason, 96 hours has become the standard for acute tests, or sometimes 48 hours for tests on invertebrates. However, acute tests lasting for 30 seconds or for 1 hour have been reported in the literature (Baker and Crapp, 1974). An assumption behind acute tests is that any substance eliciting a rapid response, such as death, at unrealistically high concentrations will be very likely to elicit effects at lower, chronic exposure concentrations. Implicit in this rationale is the idea that the ranking of the acute effects of different toxicants will be reasonably concordant with the ranking of the chronic effects. However, if time lags occur before the appearance of toxic effects, acute tests can provide misleading results. Several studies of the effects of an oil emulsifier applied to a number of invertebrate species found that, although animals survived an acute exposure and appeared to have recovered, they died a few weeks later (Baker and Crapp, 1974 and references therein). Bryan (1980) recognized that the LC_{50} approach has provided a wealth of information about the effects of metals on different species and has revealed many biological, chemical and environmental factors which modify metal toxicity. However, Ernst (1980) concluded that, for highly persistent substances having low water solubilities and high bioconcentration potentials (many organic chemicals), the LC_{50} test can provide only marginal information. 'Toxicologically it is inadequate and many biological effects of pollutants may not be evident', Moriarty (1983) argued, 'There is no good reason to suppose that there is a constant relationship, for different pollutants or different species, between the dose needed to kill and that needed to impair an organism.'

4.3.2 Acute toxicity: extrapolating to chronic toxicity

At least within species, some general relationships between acute and chronic effects are apparent. For example, Sloof *et al.* (1986) examined 164 compounds that included pesticides, organics and inorganics, and found a linear relationship between the $\log_{10}$ acute response (independent variable) and the $\log_{10}$

chronic response (dependent variable) for the sensitivities of various daphnid and fish species over approximately a 100 million-fold change in acute toxicity concentrations ($\log_{10}$[chronic] = 0.95 $\log_{10}$[acute] - 1.28, $n = 164$, $R^2 = 0.79$). This relationship explains almost 80% of the variability in the logarithm of the chronic response. At first sight this relationship seems extremely good and can in fact be quite useful if applied with care. Because the promise of extrapolation from acute to chronic response to toxicants has been eagerly received by environmental regulators and industry, and because it could potentially provide the basis for future risk assessment strategies, we will examine this relationship in more detail.

The Sloof relationship between acute and chronic toxicities is an empirical one and predicts a 19-fold greater change in the acute relative to chronic response on a log–log scale. Thus the predicted chronic response is relatively insensitive to quite large changes in the acute response. The fact that quite a wide range of acute responses predicts very similar chronic responses indicates a large degree of decoupling between these two measures of toxicity. The usefulness of the predictive equation will depend on its intended application and on the accuracy one requires. For example, the vertical range of points corresponding to a given acute toxicity value is approximately three orders of magnitude (Sloof *et al.*, 1986, their Figure 3). This means that at the lowest measured values of acute toxicity, the NOEC data values range between 10^{-5} and 10^{-2} $\mu g\ l^{-1}$ while at the upper end of the regression the NOEC range is approximately 10–10 000 $\mu g\ l^{-1}$. Because the log transformations have homogenized the variance about the linear regression line, the data points cluster about the fitted line in a reasonably homogeneous band. As one would expect, this band of data points is almost entirely enclosed by the calculated 95% prediction limits which Sloof *et al.* (1986) obtained analytically (see Sokal and Rohlf, 1981 and/or Woonacott and Woonacott, 1990 for good discussions of prediction limits and their interpretation). The prediction limits are interpreted as enclosing a region in which, given a known acute response, the probability that a single measured chronic response will fall within them is 0.95 (Woonacott and Woonacott, 1990, p. 385). When we examine the prediction limits more closely, we note that while the R^2 value is fairly high, there is still quite a large amount of variation. We can see this clearly if we back-transform the equation and examine it graphically. Transforming the log–log linear Sloof equation into a power curve we obtain:

$$\text{NOEC} = 0.052\ \text{ACUTE}^{0.95}$$

In Figure 4.1a we have plotted the predictive power relationship and the pair of lines delimiting the region within which approximately 95% of the NOEC values would lie according to Sloof *et al.* (1986). Because we did not have access to the original data, we have simply estimated the prediction limits from their plot (their Figure 3). We have drawn a graph of these limiting upper and lower lines and marked the beginning and end with open squares for clarity. Note that at this level of graphical resolution, the lower limit is coincident with the acute response axis (Figure 4.1a). Note further that the upper 95% prediction limit is very close to the line $y = x$ because of the power curve exponent near unity. Two additional aspects of this data set are also worth examining more closely. First, greater than 90% of the measured acute responses fell below 100 μg l^{-1}. This means that, had we plotted the original data, only 16 distinct points out of the 164 would be visible, spread thinly between the prediction limits. The remaining 148 data pairs would be packed into the lower left-hand corner within and closely adjacent to the square at the origin (Figure 4.1a). This can be seen more clearly if we truncate the chronic response axis at 500 μg l^{-1} and note that approximately 90% of the data pairs used to formulate the predictive equation fall within the small cross-hatched area at the lower left of Figure 4.1b.

Thus we must conclude that our ability to predict chronic responses over an extremely wide range of acute values (100 and 10 000 μg l^{-1}) is based on very sparsely distributed data (only 16 out of a total of 164 points). The log–log transformation of the data, while required in this instance to homogenize the data before least squares regression analysis, tends to obscure this fact. In addition, the predictive ability of this relationship is actually quite poor at the higher values of acute response. Given an $L(E)C_{50}$ of 10 000 μg l^{-1} there is a probability of approximately 95% that a newly measured chronic response will fall between 10 and 10 000 μg l^{-1} (Sloof *et al.*, 1986). Clearly there is room for improvement in our understanding of the relationship between acute and chronic toxic response. Improved predictions of this relationship must depend on greater understanding of the underlying mechanisms of toxicant action. The extent to which there are general mechanisms linking acute to chronic responses is almost totally unexplored. We suggest that intensive studies along these lines would well repay the effort involved.

We may be comforted by the fact that in terms of absolute values, the predictability gets worse as the acute toxicity decreases and not vice versa (Figure 4.1a). Neverthess our relative error of prediction will be the same. This can again be seen clearly by examining Figure 4.1a and noting that the probability of an additional chronic value falling between the lines is approximately 0.95 for any given acute value. This is an extremely wide range which effectively covers all values below the line: chronic response = acute response.

Because of the relative uncertainty of the relationships between acute and chronic (or higher-level) effects, those substances that elicit acute responses at relatively low doses (that are deemed to be highly toxic) are usually required to be tested under chronic exposure tests in many risk assessment protocols (Environmental Protection Agency, USA; Organization for Economic Development, France; Nordic Council of Ministers). In contrast, substances having low acute toxicity are generally not required to undergo chronic testing unless they also are slowly biodegradable or have a high potential for bioaccumulation. We can only guess how often substances judged nontoxic in acute tests

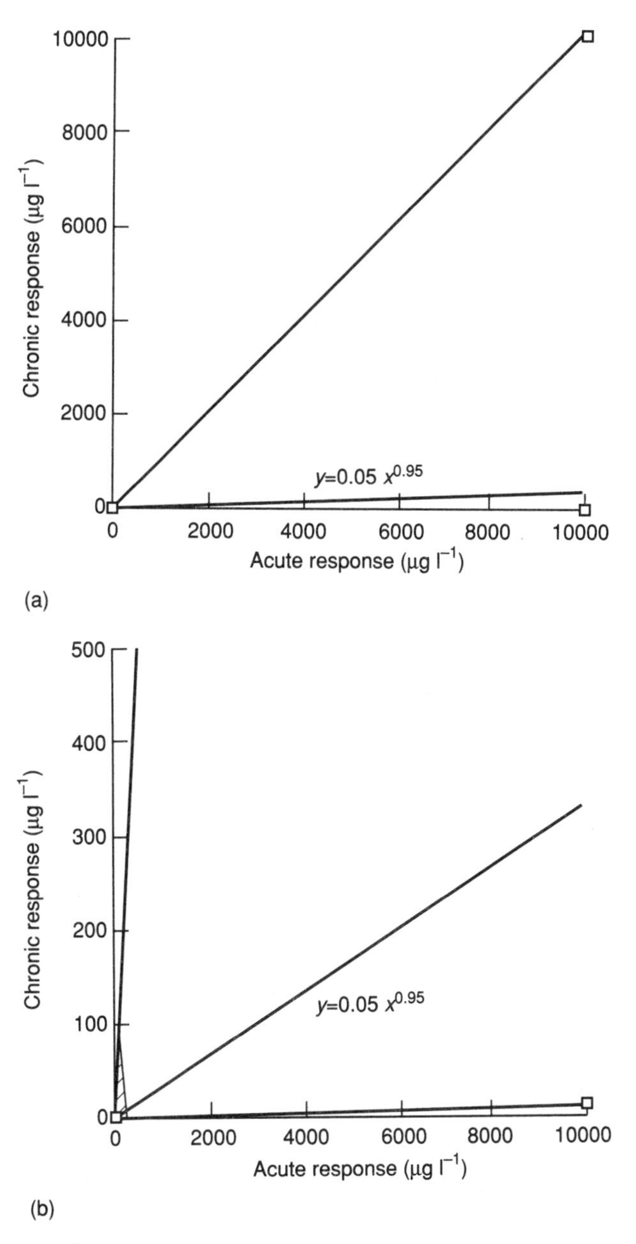

Figure 4.1 (a)Plot of the ordinary least squares regression line and associated 95% prediction limits for the relationship between acute and chronic responses to a number of organic and inorganic compounds (n for regression = 164). The upper and lower prediction limits begin and terminate with open squares. Original data from Sloof *et al.* (1986). (b)Plot of the ordinary least squares regression line and associated 95% prediction limits for chronic toxicity as a function of acute response. Note that the chronic response axis has been truncated at 500 µg l^{-1}. See text for equation and details.

will turn out to have important chronic effects, but clearly predictions based on acute toxicity can lead to serious underestimations of chronic response. A classic example of this effect is the toxicity of DDT to bird species, especially large predators such as the osprey. Although the acute effects are relatively minor, chronic exposure and bioaccumulation leads to eggshell thinning and reproductive failure. It is largely the result of extensive research into substances such as DDT that bioaccumulation potential and biodegradability have been incorporated into risk assessment protocols. However, there are some chemicals for which even the combination of acute toxicity and bioaccumulation potential may not be sufficient to predict chronic response. For example, some of the organophosphorous insecticides exert their primary chronic effects, not by bioaccumulation to high body loads, but by irreversible damage occurring during repeated periods of acute exposure (Coppage and Matthews, 1974).

4.3.3 Homogeneity of test population

Another parameter that is often considered in attempts to optimize the design of toxicity test systems is the regression coefficient (the slope) of the relationship between dose (or concentration) and response (Finney, 1978). For a log-normal tolerance distribution, the regression coefficient is equal to the reciprocal of the standard deviation of the tolerance distribution. The slope measures the variability of the test population as a function of the toxicant concentration or dose. Finney considered the slope to be a property of the population of organisms tested that is unaffected by experimental design. One's ability to estimate this slope reliably will of course be heavily dependent on the experimental design. He recommended that organisms be selected for homogeneity with respect to tolerance in order to reduce the variance of the tolerance distribution. In an earlier paper, Gaddum (1933) advocated the use of inbred Wistar rats because they showed higher regression coefficients than a genetically mixed group. However, Finney (1978) noted, 'a highly inbred line of animals can be undesirable, because of reduced vigor, and possibly increased variability in some physiological responses.'

The benefits involved with increasing the homogeneity of test organisms are obviously closely linked to the purpose of the test. For example, use of inbred strains of rats might be highly desirable if we were interested in determining the effect of a single gene substitution on the acquisition of resistance. However, if the purpose of a test is to assess the risk that a potential toxicant or pollutant will have a deleterious impact if released into the environment, or to rank the relative toxicity of different chemicals, then limiting the genetic heterogeneity of test organisms is a misplaced priority (Forbes and Depledge, 1992a).

The problem of ranking chemicals based on data obtained from a single clone is illustrated by the following example. Several recent studies have found significant differences among clones of *Daphnia magna* in tolerance to a range of chemical toxicants (Baird *et al.*, 1989; 1990). The 1990 study examined the toxicities of cadmium and dichloroaniline (DCA) to different clones of *Daphnia magna* by comparing means and 95% confidence limits of the LC_{50} tests. The clones differed markedly in their susceptibility to the toxicants, and among-clone variation was greater for cadmium than for DCA. Thus the ranking of clones, in terms of sensitivity, differed for the two toxicants. Clones that were very sensitive to cadmium were not necessarily sensitive to DCA and vice versa. Ecological effects aside, these results need to be taken into consideration by those in regulatory agencies involved in the classification and ranking of chemical substances because the relative ranking of chemical toxicities depends on which clones are tested.

4.3.4 *Acute toxicity: other limitations*

Kinne (1980) argued that identification of pollutant impact on natural populations of organisms is complicated by various compensation mechanisms that tend to mask pollutant effects and result in time lags between initial impact and observable effects. In this regard, acute mortality tests, which are designed to provide extreme conditions, prevent, according to Kinne, the activation of compensatory mechanisms and thereby greatly simplify the relation of cause and effect (although time lags can still be a problem). Although acute tests may provide a quick general assessment of pollutant toxicity, death has been widely criticized as being a crude endpoint. In addition, the duration of such tests is too short to adequately assess ecologically relevant consequences of pollutant exposure. 'Ecologically, the death of an individual is largely irrelevant. What counts here most are changes in competitive performance of coexisting groups of organisms and flow patterns of energy and matter at the ecosystem level' (Kinne, 1980). It is widely recognized that sublethal effects on individuals may be as important to the continued success of the population as lethal effects.

Stern and Walker (1978) made the important point that the LC_{50} may be valid only for the test population, but not the test species as a whole, and that spatially and temporally varying populations of a species should be tested if the purpose is to estimate the entire species tolerance distribution. They recommended a thorough investigation of the pollutant's properties and environmental characteristics along with information regarding the sensitivity of different life stages of organisms that would be exposed to the pollutant. They proposed that information on the 0% and 100% mortality concentrations, slope values for mortality curves, information on sublethal responses and any other relevant data be reported along with the LC_{50} results.

Blanck *et al.* (1984) tested freshwater algal species for tolerance to 19 different chemical compounds (Table 6.1) and found that species-dependent variation in sensitivity was up to at least three orders of magnitude, and more recent work brings this estimate up to 10^6, depending on the specific chemical tested (H. Blanck, personal communication). They were unable to identify species that were generally sensitive or generally tolerant to a wide range of chemicals and recommended that algal test batteries replace single-species test systems. Similarly, Seager (1988) summarized literature data on the toxicities of the insecticide permethrin to freshwater species and tributyltin to marine and estuarine species. Reported lethal concentrations for permethrin in rainbow trout spanned three orders of magnitude. Two sets of tests of the acute toxicity of copper to rainbow trout were performed in the same laboratory, using the same stock of fish and the same water source (discussed in Sprague, 1985). The LC_{50} values, which differed by a factor of 5.5, both appeared to be valid.

Sublethal effects of tributyltin on oyster (*Crassostrea gigas*) spat occurred at concentrations as low as 0.01 μg l^{-1} whereas LC_{50} values for adults were reported to be 1800 μg l^{-1} (Seager, 1988). This fivefold range in effect level highlights the importance of life stage, methodological variability and choice of endpoint in accurately assessing toxicity. It ignores the issue of species interactions, which would probably increase the variance in response even further. Another important conclusion from Seager's analysis is that the variability in toxicity within species can be equal to or greater than the variability among species. For example, the variability in toxicity of permethrin within the genus, *Daphnia*, is greater than the variability among all insect species for which data were compiled.

Giddings (1986) concluded that the standard single-species laboratory bioassay will continue to be the most widely used method for evaluating chemical hazards. 'It is unlikely that industries, regulators, or the courts will abandon methods and interpretations that have evolved over the years and are now widely accepted.' As ecotoxicologists continue to debate the merits and flaws of the $L(E)C_{50}$ test, decisions about the safety of substances released into our environment must continue to be made.

4.4 Problems with standardization

> Since all models are wrong the scientist must be alert to what is importantly wrong. It is inappropriate to be concerned about mice when there are tigers abroad.
> G.E.P. Box (1976)

> If an experiment is not worth doing, it is not worth doing well.
> Sir Peter Medawar (1979)

4.4.1 Methodological and statistical strategies

Keeping the debate on ecological relevance and the caveat of Sir Peter firmly in mind, there are nevertheless several important ways in which laboratory toxicity tests can be improved. Perhaps the most important of these concerns proper experimental design and choice of statistical methods. Like microscopes, pH meters, scintillation counters and any other equipment that we use, statistical analysis is a tool that facilitates our ability to make measurements and interpret results. Although decisions concerning the choice of laboratory equipment used in an experiment are most often made well before measurements are begun, the choice of statistical analyses is sometimes left until the experiment is completed. This is too late. Consideration of statistical analyses should begin during the design phase of an experiment, well before any data are collected. Although statistical analysis can greatly facilitate our ability to detect and interpret results, especially when differences among treatment groups are not large, nothing can rescue a badly designed experiment.

The recent renewed attempt to come up with improved guidelines for the standardization of ecotoxicological testing methods among EC member countries is currently focusing almost exclusively on laboratory practice and the actual performance of the tests. The implication is that analytical/statistical problems associated with the design and interpretation of toxicological tests have been solved and that tests are interpreted in an optimal and reproducible manner. This is simply not the case. Although statistical theory exists to handle most of the analyses, it is scattered in a literature generally unknown and inaccessible to practising ecotoxicologists. In current guidelines, statistical recommendations are given without adequate reference and sometimes appear to be rather arbitrary. For example, 'At least 20 animals, preferably divided into four groups of five animals each, should be used at each test concentration and for the controls.' (Organization for Economic Cooperation and Development, 1981, guidelines for the *Daphnia* sp. acute immobilization test). The references cited in support of these recommendations do little to explain the rationale behind the bald statement. This is inexcusable. There is no explanation given for the recommendation of 20 animals, thus readers are not provided with any information to enable them to decide whether 20 animals will provide sufficient replication. Such a decision is likely to be critically dependent on the purpose of the experiment, the inherent variability of the test population, the chemical to be tested and the desired level of confidence in the results.

It is critically important that the analytical procedures be given as much attention as the experimental measurements. Investigators can make informed choices about experimental design only when equipped with a basic understanding of the analytical methods themselves. Without proper analysis, much of the time and effort of standardization is wasted. To this end, a basic understanding of the analytical and statistical methods of proper data analysis is essential for all those involved in conducting and interpreting tests. For example, the usefulness of simple pilot experiments for acute (and other) tests has been neglected in the past. Pilot experiments allow the choice of concentrations in an acute test to be closely clustered near the expected LC_{50}, dramatically increasing one's confidence in the estimate. In addition, variance of the tolerance distribution is a function of both the test population and the chemical tested (e.g., Baird *et al.*, 1990; Baird *et al.*, 1991). As the variance in tolerance influences the reliability of median lethal (or effect) estimates, the use of a single, constant number of test organisms for all chemicals may not be the most rational approach. To maintain a more consistent degree of confidence in our tolerance estimates, we should base the number of test animals for given chemicals on the variance of the tolerance distribution, hence the need for pilot experiments.

4.4.2 Precision and accuracy

To examine the nature of experimental precision and accuracy in greater detail, we focus, as in our earlier examples, on the acute concentration–response relationship as a model experimental system. First, however, it is important to distinguish between these two properties. Ideally a toxicity test should be both precise and accurate. Precision is inversely related to the spread or dispersion among replicates. The greater the precision of a test, the greater its reproducibility or reliability. This is distinct from the concept of accuracy, which refers to a lack of bias. The accuracy of a test is its ability to estimate the 'truth' on average. We will be concerned here with biased test results due to violations of the underlying statistical assumptions. It is very important to keep the distinction between accuracy and precision in mind because the specific conditions and assumptions under which a test is performed can cause accuracy and precision to vary independently. This means that we can have a very precise but biased test, leading to a false sense of security regarding the quality of our results. For a standard LC_{50} test, accuracy can only be evaluated by critical examination of the test assumptions and an analysis of how well the test results conform to the theoretical model (see Forbes, 1993 for a more detailed discussion).

4.4.3 Hypothetical example

A experimental simulation may help to illustrate the problem. Suppose we would like to measure the acute effect of atrazine on *Daphnia magna* by conducting a standard, 24-hour EC_{50} acute immobilization test using the method of probit analysis (e.g., Organization for Economic Cooperation and Development, 1981). We decide on a control and six geometrically increasing treatment concentrations centred roughly on what we believe to be the true EC_{50} for this population of *D. magna*. We have not performed any preliminary experiments, and there are no data on the effect of atrazine on *D. magna*, so we estimate the true EC_{50} for atrazine using published literature values for other *Daphnia* species. We have used a geometric treatment series where each treatment concentration is double the previous one so that the observations are evenly spaced when the concentrations have been log-transformed (Table 4.1). We have selected the treatment concentrations such that our literature estimate of the EC_{50} is centred within their range. Following the OECD

Table 4.1 Atrazine concentrations for the simulated *Daphnia magna* acute immobilization experiments described in the text. TP_i is the true or actual proportion of the test population shown in Figure 4.2 that will respond to a given atrazine concentration.

Atrazine ($mg.l^{-1}$)	log_{10}[Atrazine] ($mg.l^{-1}$)	TP_i
0.500	0.301	0.299
1.000	0.000	0.509
2.000	0.301	0.620
4.000	0.602	0.728
8.000	0.903	0.809
16.000	1.204	0.877

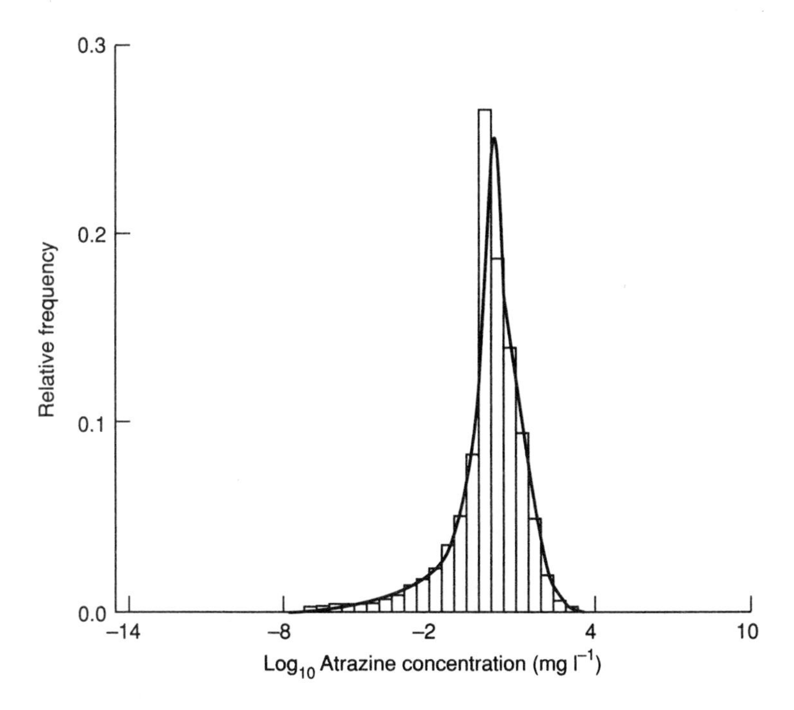

Figure 4.2 Tolerance distribution for the simulated acute immobilization response of *Daphnia magna* to atrazine. The 'total population size' is equal to 5000. Random samples were taken from this distribution at each experimental treatment concentration and used to generate the probit values shown in Figures 4.3 and 4.4. True mean μ_T = 0.766 mg l^{-1}, standard deviation σ_T = 28.119 mg l^{-1} and median = 0.999 mg l^{-1}. The text has a more detailed description of the procedure.

guidelines we decide to allot five animals to each of four replicates per treatment, using a total of six treatments and one control. This gives us a total of 140 animals when we include the control. This design conforms to the present guidelines for the conduct of an acute *Daphnia* immobilization test. Further, suppose that the true tolerance distribution of the response of *D. magna* to atrazine is not log-normal. Because of the mechanism underlying the effect, the true distribution is skewed left after log-transformation, not normalized as is assumed in probit analysis. Because this is a simulation we can know exactly the true, underlying distribution of the response of *Daphnia magna* to atrazine. This distribution is shown in Figure 4.2.

We can now investigate how specific test conditions and departures from assumptions can affect the results of our analysis. It should be clear that the skewed nature of the tolerance distribution will cause the true mean and median to differ from each other in violation of the assumptions of the analysis. The mean will be somewhat lower than the median, pulled left by a few very sensitive animals (Figure 4.2). Further, the standard deviation will probably be different from that of a normal distribution with the same mean (or median).

Two important points must be made concerning the statistical nature of the response for each replicate. First, an individual *Daphnia* is either immobilized or it is not, making this a quantal-response experiment by definition. Second, given the number of animals used for each replicate (five), there are only six possible outcomes. All animals may be immobilized, none may be, or some fraction in between. This means that the measured experimental outcomes will be expected to follow the binomial distribution. Thus there will be a unique binomial distribution for each treatment concentration, the exact nature of which is determined by the number of animals used per replicate and by the true probability (P) of immobilization (which we know because we have constructed the 'true', underlying distribution) (Table 4.1). We can now run a *Daphnia* sp. acute immobilization experiment by randomly sampling the binomial distribution for each value of P in each replicate and calculating or looking up the probit values for the percentage of affected animals in each replicate. We then plot the results of the experiment in Figure 4.3. To simplify the analysis we assume that there was no mortality in any of the controls.

In this illustrative example, we will analyse the data in a cursory fashion, calculating only an ordinary least squares regression line (OLS) and associated 95% confidence bands (Figure 4.3). These will only approximate the more appropriate solution that would have been obtained using weighted regression or maximum likelihood methods. Our solution is only an approximate one because of two features inherent in probit analyses of quantal-response data. First, the response variable, immobilization, is distributed binomially, not normally as is assumed for OLS (Draper and Smith, 1981). Second, measurement error is distributed about the centre of the tolerance distribution such that the data need to be weighted accordingly (Bliss, 1935; Finney, 1971; 1978; Forbes, 1993). Nevertheless, we will continue with the example realizing that

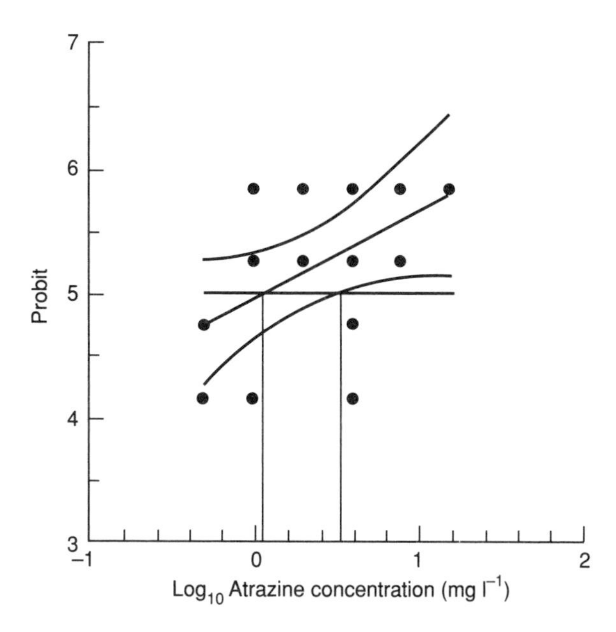

Figure 4.3 Plot of the probit regression for the first *Daphnia* immobilization simulation. The biconcave curves are the 95% confidence bands of the regression. For this simulation, five *D. magna* were used for each replicate. See text for discussion and details.

we are only approximating a more complete analysis. Readers are encouraged to consult Finney (1971; 1978) and Forbes (1993) for more detailed discussion and examples of the more complete methods that should be used in all serious analyses.

4.4.4 Short theoretical digression

In this example, the 0% and 100% effect levels will not be included in the analysis. In the log-normal tolerance distribution theory used in probit analysis, the true 0% and 100% effect levels are plus or minus infinity. Thus probits cannot be given in the standard tables. This means that probit values for these effect levels must be estimated. In general, the fewer animals used per replicate, the greater the likelihood of obtaining a 0% or 100% effect-level result. For example, when the actual percentage immobilization of *Daphnia* is near 100%, small sample sizes often cannot detect the tiny fraction of animals that are unaffected. Although there are methods for estimating probits for these extreme values, they require certain assumptions that we would like to avoid making at this time (see appendix by R.A. Fisher in Bliss, 1935 for a detailed discussion).

4.4.5 Hypothetical example revisited

Several important features of Figure 4.3 are worth noting. The first is that because only five animals were used in each replicate, only four probit values are possible along the *y*-axis of the graph when the 0% and 100% effect levels are discarded. This can lead to the very coarse-grained, low-resolution plot shown in the figure. Note that all four possible probit values have been measured for $\log_{10}$ ([atrazine]) = 0.6. Also, random sampling produced an identical probit value of 5.84 for five of the six treatment concentrations. For this particular experimental run, it would be virtually impossible to draw a reasonable line 'by eye'. Note also that due to the small number of *Daphnia* per replicate and the nature of the sampling process, five out of a possible 24 replicates had measured 0% or 100% effects, leaving only 19 points on which to base the initial regression line. This amounts to greater than a 20% loss in the original replication. We should also note that because this is a simulation, the variability of the response at a given toxicant concentration is as low as possible under the assumption of a binomially distributed immobilization response. Real toxicity test results can be expected to contain some additional experimental error.

Because we expect the probit concentration to straighten out the plot in Figure 4.3, we calculate an OLS regression line and find it to be

$$\text{probit} = 0.679 \cdot \text{LOG [atrazine]} + 4.965$$

Further, we calculate an *F* ratio of 5.67 and, assuming that all the assumptions for OLS hold (although we know they do not), we find that the probability of obtaining a value this large, if the true regression slope is really zero, is only about 0.03 (Draper and Smith, 1981). To calculate our estimate of the EC_{50}, we set probit equal to 5.0 and rearrange the above equation to get $EC_{50} = 0.50$. Taking the antilogarithm we have $EC_{50} = 1.122$ mg l^{-1}. Thus we expect half of the *D. magna* from the population we are trying to estimate to be immobilized by an atrazine concentration of 1.122 mg l^{-1}. The standard deviation of the assumed log-normal tolerance distribution is the reciprocal of the regression slope, so that our estimate of the standard deviation for the tolerance distribution (SD_T) is (Finney, 1978)

$$SD_T = 1/0.679 = 1.473$$

We assumed the log-normal tolerance distribution so that the estimated mean and median are necessarily equal.

How do these estimates compare with the true parameters for the hypothetical *D. magna* population? Inspection of the caption for Figure 4.2 reveals that the true mean (μ_T) and standard deviation (σ_T) of the tolerance distribution are -0.116 and 1.449 respectively. Taking the antilog of -0.116 we have 0.766 mg l^{-1} as the true value for the mean and antilog (-0.00055) = 0.999 as the true value for the median or EC_{50}. Thus our estimated EC_{50} of 1.122 was in error by about 12% (% error = {1.22 - 0.999]/0.999}·100 = 12.3%). The true mean was even less accurately estimated because of the degree of skewness in the tolerance distribution. Our estimate of the standard deviation of atrazine tolerance in *D. magna* fared much better. The error was less than 2%.

In order to obtain confidence limits for our estimated EC_{50}, we need to employ the method of inverse prediction. The calculations are somewhat more complicated than the standard ones (Draper and Smith, 1981; Sokal and Rohlf, 1981 give detailed description of the calculations). However, a graphical method can be used to obtain approximate limits (Draper and Smith, 1981). The graphical approach involves drawing a horizontal line at the probit value we wish to use to predict the median effect concentration. In this example we wish to predict the atrazine concentration corresponding to the probit value of five. The predicted EC_{50} is found by dropping a perpendicular line from the point of intersection of the horizontal and the fitted regression lines. The approximate upper (X_u) and lower (X_l) 95% confidence limits are found by dropping perpendiculars from the intersection of the horizontal line and the lower and upper 95% confidence bands respectively (Figure 4.3) (Draper and Smith, 1981). In this case we are faced with one of the peculiarities that can arise when the regression is not well determined. The upper confidence band flares upward at the point of interest such that there is no intersection which would allow us to determine a lower confidence limit for the EC_{50} (Figure 4.3). This suggests that the solution to the quadratic equation for X_l is a complex number and no real number solution is therefore possible. (Draper and Smith, 1981, p. 49). In fact, inspection of Figure 4.3 reveals that it would be impossible to draw any horizontal line (that is, use any probit value as a predictor variable) that would intersect both the upper and lower confidence bands, that is, allow estimation of both X_l and X_u.

4.4.6 Improving the estimates

What happens when we redesign the experiment so that we improve the precision of the estimated EC_{50}? One way we can improve precision is by including more *Daphnia* in each replicate (Forbes, 1993). This strategy will have two effects. First, our estimates of P for each replicate will be more reliable, which will in turn increase our confidence in the estimated tolerance distribution parameters. Second, we will be less likely to get 0% and 100% effect levels, thereby increasing the number of replicates we can use to calculate the OLS line without having to estimate probit values. For this experiment, we will use 15 *D. magna* per replicate. All other conditions and assumptions are the same as in the previous experiment. This means that including controls we will use 15 (animals/replicate) × four (replicates/treatment) × seven (treatments) = 420 animals in total. This is a substantial increase in effort and probably expense. Is it worth it?

When we perform this experiment, and look up the probit values for the percentage of *Daphnia* affected, we obtain the results shown in Figure 4.4. Overall, the scatterplot looks like a fairly reasonable result from a typical concentration–response experiment with replication. Notice that because there are now 16 possible binomial outcomes for each value of P, there is considerably less graininess or discreteness in the distribution of values along the y-axis (Figure 4.4). Furthermore, it might be possible to draw a fairly reasonable 'eye fit' to the data. Comparison of Figures 4.3 and 4.4 reveals two quite different scatterplots. Remember that the underlying tolerance distribution for the response of *D. magna* to atrazine is exactly the same, only the sampling design has changed. The OLS fitted line for this experiment is

$$\text{probit} = 0.891 \cdot \text{LOG}\,[\text{atrazine}] + 4.892$$

With 15 *D. magna* per replicate, we are able to get graphical estimates (though just barely) of approximately -0.28 and 0.36 for X_l and X_u respectively (Figure 4.4). Notice that if we had centred the regression line more closely on the EC_{50} we would have obtained tighter 95% confidence limits. Our revised, theoretically more precise, estimate of the variability of the *Daphnia* tolerance distribution is now

$$SD_T = 1/0.891 = 1.122$$

Solving the regression equation for the new estimate of the EC_{50} we have $EC_{50} = \text{antilog}(0.121) = 1.322$ mg l^{-1}. Comparing our estimated EC_{50} with the true median we find that it is now too high by more than 30%. Our estimate of the spread of the tolerance distribution got worse as well. We have underestimated σ_T by almost 9%. This was supposed to be an improvement in our experimental design. What happened? The improved precision has clearly been decoupled from the accuracy of our estimates.

Simply put, because of the violations of the assumptions of our analysis, we are getting better at estimating the wrong thing. We have assumed that the probit transformation would straighten out the data and that the fitted regression line would allow us to accurately estimate the spread and location of the *Daphnia* tolerance distribution. However, we know that these assumptions are not valid. The effect can be seen graphically.

The data points in Figure 4.5 are simply the experimentally determined points for the two previous simulation experiments that were illustrated in Figures 4.3 and 4.4. Points are plotted for both five (open squares) and 15 (filled circles) animals per replicate. Superimposed on the basic scatterplot we have added two additional curves. The smooth curve was determined in the following manner. We can calculate from the simulated distribution in Figure 4.2 the actual proportion of individuals that will be immobilized by a given atrazine concentration. We will call this the 'true proportion' immobilized or TP_i. These values are tabulated in Table 4.1. Following the normal procedure for a probit analysis we assume that this distribution is in fact Gaussian or normal. We then ask the following question. If the distribution were indeed normal, what

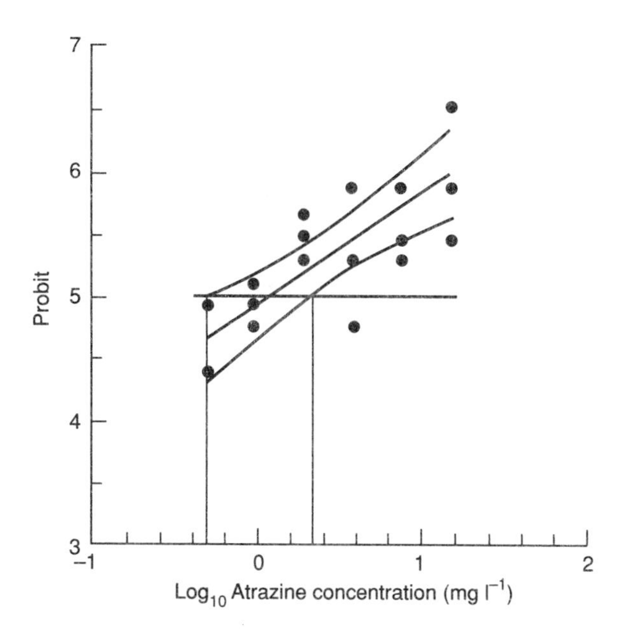

Figure 4.4 Plot of the probit regression for the second *Daphnia* immobilization simulation. For this simulation, 15 *D. magna* were used for each replicate. See text and the caption of Figure 4.3 for details.

is the expected fraction of organisms affected (P_i)? We then use TP_i to calculate probit values in the usual way (i.e., $TP_i \rightarrow P_i \rightarrow Probit_i$). These probit values are plotted as the smooth curve in Figure 4.5. The important point to note here is that this curve is the centre of our binomial sampling distribution for each treatment concentration. This means that the experimentally determined probit values will tend to fluctuate randomly about the curved line. The 'squiggly line' was calculated differently. For this line the actual fraction of organisms affected was used to obtain a probit value directly ($TP_i \rightarrow Probit_i$). Thus the squiggly line represents the real (skewed) distribution.

We can now see on the graph what has happened. Increasing the precision of our sampling design has improved our ability to converge on the wrong curve (Figure 4.5). That is, we are improving our estimate of the smooth curve in Figure 4.5. Increased effort to improve our precision would only aggravate the situation. Note also that there is an inherent 'lack of fit' error component because we are fitting a straight line to a curved function (Draper and Smith, 1981; Forbes, 1993). Thus, while precision can be controlled by good laboratory practice, choice of clones and other experimental design factors, it is critical to remember the importance of accuracy as well, and that accuracy can vary independently of precision. The accuracy of the test results depends on the assumptions of the analysis and on how well the data fit the underlying model.

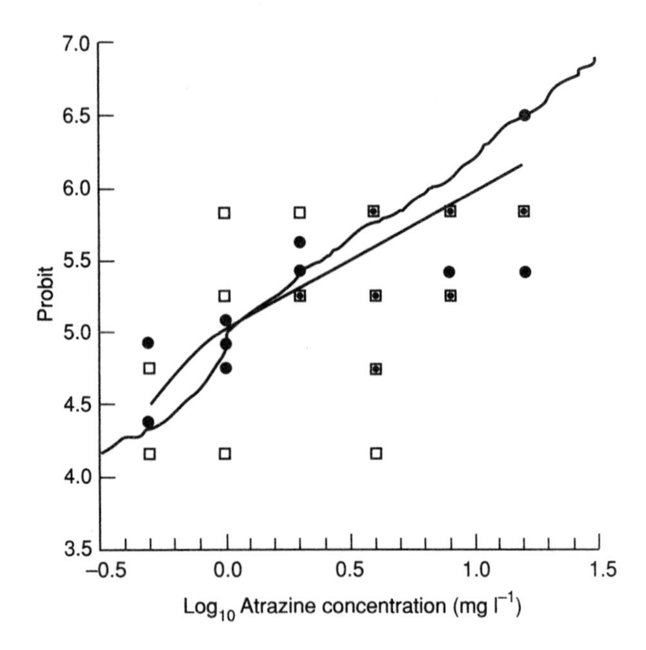

Figure 4.5 Probit plots for the first and second *Daphnia* immobilization simulations. (□) = First simulation, five animals per replicate; (●) = second simulation, 15 animals per replicate; ⊡ shows points that overlap. See text for an explanation of the two superimposed lines and a discussion of their significance.

4.4.7 Need for reliable data

Why all the fuss? If the estimated EC_{50} is going to be multiplied by a factor of 1000 for regulatory or management purposes, why worry about a relatively small 30% error? Certainly if the only purpose of an analysis is to obtain a crude estimate of acute toxicity, it may be reasonable to carry out a non-rigourous analysis – or is it? Referring back to Figure 4.3, recall that we were unable to get an estimate of the 95% confidence limits of our EC_{50}, even in an analysis that followed the current OECD guidelines to the letter. This means that for the first experiment we have no statistical estimate of the reliability of the estimated EC_{50} at all. Thus we have no basis for determining whether or not an application factor of 10, 100, or even 1000 is appropriate. We cannot estimate the odds that we will obtain an EC_{50} 100 times higher (or lower) if we were to repeat the experiment. These considerations bear directly on the fact that regulatory decisions or risk identification/assessment protocols are often based on simple numerical standards, such as the use of ratios between no effect and predicted environmental concentrations for the setting of maximum concentration limits. Clearly, when single numerical values are used as some form of cutoff criterion, the measurements bearing on them must be made as reliable as possible.

An additional important reason for pursuing more rigour in the statistical analysis of toxicity and pollution data concerns the ability of regulatory decisions based on toxicity tests or experimental results to stand up in a court of law. Results on which economically influential regulations or decisions will be based can be expected to undergo intense scrutiny by the parties directly involved and must be able to stand up to the investigation. If the studies and the regulations based on them are not statistically sound, they risk being invalidated or overturned.

A case in point is illustrated by the recent controversy surrounding a study of the initial effects of drilling-related pollution around the Ekofisk oilfields in the North Sea (Gray *et al.*, 1990). Contemporary reviews of the effects of oil have generally concluded that benthic communities in the North Sea are influenced by pollution emanating from the drilling platforms out to approximately 1 km. More severe effects are generally conceded to occur within 500 m of the platform (Hartley, 1984; Kingston, 1987). Gray *et al.* (1990) re-analysed data from the Ekofisk field in the Norwegian sector from an original survey by Hobbs (1987). In analysing this data, Gray and colleagues employed a more sophisticated and powerful multivariate analysis. The newer results revealed a clear signal of the platform's effects on benthic communities that could be detected at a distance of up to 2–3 km rather than 1 km as was previously believed. This conclusion has been hotly debated because simple geometric considerations reveal that a much greater fraction of the North Sea benthos than previously thought is influenced by the drilling activity. If the oil-field effects extend out to distances of 2–3 km, the area of North Sea bottom influenced by drilling is 5–9 times greater than estimated previously (Gray, personal communication). These results have the potential to significantly influence future regulatory decisions regarding drilling activity in the North Sea. This provoked a strong response by some oil companies to discredit the analysis and invalidate the resulting conclusions. One company even went as far as employing two statisticians whose sole job was to find flaws in the analysis (Gray, personal communication). Interestingly, the argument that the drilling effects do not extend out to 2–3 km is based on statistically invalid grounds. It is dependent on the conclusion that failing to detect an effect means that there is in fact no effect. Without a statistical power analysis we cannot know the probability that this conclusion is correct. Clearly, the results of analyses that will be used as a basis for environmental legislation must be able to withstand this kind of close scrutiny. An EC_{50} value with no accompanying estimate of reliability does not meet this minimum standard.

4.5 Ring tests, variability and standardization

4.5.1 Round-robin or ring tests

The purpose of ring testing is to ensure that laboratories in different agencies, states or countries meet the same high standards of scientific quality (Rand

and Petrocelli, 1985). These tests require that a special design protocol be used for the specific purpose of assuring laboratory quality control. Otherwise valid comparisons among laboratories will not be possible. Recent discussions have focused on the fact that interlaboratory calibration exercises have often failed to reach desired levels of concordance among the participating laboratories. A number of factors have been implicated in causing variability in the results, including diet, culture medium, health of the organism, purity of the test chemical and certain inherent characteristics of the test chemical (Cowgill, 1986).

Lee *et al.* (1986) noted that, although test methods for acute and chronic toxicity tests have been standardized at the national and international level, this standardization has not extended to the culture methods of test organisms. They pointed out that although test results are considered valid only when certain survival and productivity rates are met in the control treatments, little advice on how to go about meeting these criteria is given, particularly with regard to culture conditions. Guidelines from both the Organization for Economic Cooperation and Development, 1984 (No. 202) and the Environmental Protection Agency environmental effects test guidelines currently allow for free choice of food species and feeding regimen. Thus in recent years, there has been a number of investigations of the effects of environmental variables, such as food regimen, on culturing success and toxicant response of common test species (Lee *et al.*, 1986 and others).

Given this multiplicity of potentially interacting factors, Cowgill (1986) recommended standardizing diet and water used to culture test organisms and assessing and controlling organism health before any test. In reference to chemical characteristics, Cowgill (1986) reported on evidence from previous test results that the 'inherent percent coefficient of variation' is related to a test chemical's water solubility, molecular weight, vapour pressure, boiling point and level of toxicity. Thus, to improve results of interlaboratory tests (that is, to reduce variability), Cowgill's (1986) recommendations included careful selection of the test chemical. 'Test compounds that are nontoxic or slightly toxic, have a high solubility in water, are of low molecular weight, have a low boiling point, a high vapour pressure, and few chlorine atoms within their structure will provide better precision than those compounds that have characteristics distinct from those just mentioned.'

Similarly, the Environmental Protection Agency has recommended the use of reference toxicants, to provide information about differences in sensitivity among organisms, as an important quality insurance practice (Jop *et al.*, 1986). They have recommended that a reference toxicant has the following characteristics (Jop *et al.*, 1986): It must have a history of use in toxicity testing, be fast acting and nonselective, have low toxicity to man, require simple analytical procedures, be nonsensitive to environmental variables, and show a close approximation between nominal and actual concentrations. Unfortunately this approach 'solves' the problem by developing test protocols that essentially have very low discriminatory power. The use of such well-behaved chemicals for ensuring methodological standardization within and among laboratories may give a false sense that standardized criteria are being met. For example, if tests with a reference toxicant provide reproducible results because the behaviour of such a substance is insensitive to environmental fluctuations, of what use is this for standardizing test systems for determining the toxic effects of the many pollutants that are critically sensitive to fluctuations in environmental variables? If it is important that laboratories maintain a precise and constant temperature, how will using a test substance whose toxicity is insensitive to temperature fluctuations ensure that equivalent standards are being maintained? More effective standardization could be achieved by either measuring the variables of interest directly or by performing interlaboratory test comparisons using substances that are very sensitive to variation in environmental parameters. One good approach to the problem would involve partitioning the variance due to the various environmental and treatment effects. This could be done using analysis of variance and is the type of information critical to the optimal design of experiments.

There have been few published sensitivity or error-partitioning analyses of the different sources of variation within and among test systems. Thus we do not know how many test systems produce data that can be used to compare the uncertainty arising from differences in test protocol with the uncertainty arising from other sources. How much does the variability among laboratories influence the calculation of safety or application factors? Answers to these questions should be obtained routinely as part of the design of toxicity-testing protocols.

4.5.2 Daphnia *variability problem*

Several of the above issues are brought into sharp focus when one examines a recent series of European workshops called to address the problem of excessive variation in test results among laboratories. A letter (dated 11 February 1991) from the OECD secretariat addressed to national co-ordinators and experts on aquatic ecotoxicology, announced a workshop to address the subject of 'sources of variation in ecotoxicological tests with *Daphnia magna*.' The letter states,

> For some time there has been an awareness that data from experiments to assess the effects of chemicals on reproduction in *Daphnia magna*, designed according to the various existing guidelines (e.g., OECD, EEC), can be **very variable** [emphasis added]. Results from recent ring-tests have demonstrated considerable variation between laboratories ... [there has] emerged a general consensus that variability in test results both within and between laboratories could only be reduced through improved standardization of the method, particularly with respect to the choice of genotype, food and culture medium used.

What is meant by 'very variable'? These concerns date back to a workshop held in 1989 on the same topic which resulted in several recommendations to standardize the protocol for the so-called *Daphnia magna* 21-day reproduction test for incorporation into the OECD Guidelines for Testing of Chemicals (OECD Guideline No. 202). This test quantifies the toxicity of a chemical by measuring the production of viable offspring by *D. magna* exposed to a range of toxicant concentrations for 21 days.

Based on the extensive previous experience of the workshop participants, it was thought that the interlaboratory variability could be reduced by standardizing three critical factors. These were the genotype of the test animals, the food ration and/or feeding protocol and the culture medium. To this end the 1989 participants designed a research programme to identify what they termed the 'optimal' conditions for carrying out the reproduction test in terms of the genotype, food ration and culture medium used. The experiments needed to determine this standardized protocol were then performed. It is important to note that this research was designed to arrive at the optimal choices for the food, genotype and medium conditions to be recommended for the test. For example, the optimal genotype was defined in terms of juvenile production rate and toxicant sensitivity. Presumably, this genotype would be the one with the highest juvenile production rate or greatest sensitivity to the reference toxicants, although this is not completely clear from the report. Other similar definitions were used regarding the optimum culture medium and food ration. It was apparently assumed by the workshop participants that this approach would lead to decreased variability in test results reported by different laboratories.

While standardizing the testing protocol may well reduce laboratory variances, this was never tested directly. Given the problem of high test variability among laboratories, it is clearly valid to standardize the test conditions so that appropriate comparisons can be made among laboratories and the variability of the test system can be analysed directly. It would have been of interest to know how much of the variability was due to the laboratories and to each of the three critical factors discussed above.

We suggest that standardization of the test protocol is only the first step in a two-part process. Once some form of adequate standard test conditions are agreed upon, one can begin to ask some specific questions. For example, one relevant question is the following. When using a single clone, how much of the total test variability in the production of surviving offspring can be attributed to the different laboratories and how much is due to variation among individual *D. magna*? This is essentially an evaluation of the homogeneity in test execution among the various participating laboratories and is the proper focus of a ring or round-robin test system. This is a test for good laboratory practice. The most straightforward and economical approach to partitioning the experimental variance in this way would probably be to employ an analysis of variance design. For example, if a single classification design were used, with the laboratories as treatments, we could then make a simple *F*-test for the significance of the laboratory effect. In addition we could obtain an estimate of the percentage of the total variation in the ring test results that was due to the different laboratories and the percentage due to error within laboratories.

But say we detect no difference among the participating laboratories. What then? We cannot validly conclude that no effect exists. To do that we would need to have some idea of the probability of making a type II error (β). Like the example of the effects of drilling for oil in the North Sea, in order to make statements concerning the probability that different laboratory results are not very different we must perform a power analysis of our experimental system (see section 4.6). This will have the positive effect of focusing attention on the size of the difference between laboratories that we are interested in detecting. We need, in effect, to decide how similar we would like the different laboratory results to be. This decision must ultimately depend on the use to which we plan to put the test results.

The above approach can of course be extended in a simple fashion to analyse the variability and effects of any additional factors that might be of interest to test developers. For example, we could examine the effect and added variability of different foods, such as *Chlorella vulgaris* vs *Selenastrum capricornutum*) or food rations within a particular clone and laboratory. This experimental programme has the twin advantages of allowing direct analysis and quantification of the variability and of forcing decisions to be made on the size of the effect deemed important.

4.5.3 Discriminating between standardization to analyse variability and standardization to measure toxic effects

One of the conclusions of the second workshop was the recommendation for use of a single clone. This recommendation of a single clone or genotype of *Daphnia* for use in the test guideline for the 21-day *Daphnia* reproduction test is an example of the failure to make the proper distinction between development of protocols for testing laboratories and protocols for testing organisms.

The central problem seems to be a failure to draw a clear distinction between the goal of a ring testing exercise and the ultimate purpose of the toxicity test. These goals are different. Ring tests should be designed to test laboratories or protocols, whereas toxicity tests are designed to provide some meaningful measure of the adverse effects of a chemical on biological systems. The danger occurs when we mistakenly apply the same kind of design rationale to both kinds of test.

4.5.4 The concentration–response curve and genetic variability

Gaddum (1933) noted that the slope of the line made by plotting the response versus the logarithm of the toxicant concentration was a direct measure of the uniformity of response of the organisms used in the experiment, and he argued that the use of closely inbred animals was highly advantageous in such tests because of their genetic uniformity. In reference to toxicity experiments employing asexual organisms, Landis (1986) stated, 'Of course, to ensure the dependability of the data, genetically identical organisms are vital.' Among other things, advocating the use of inbred or clonal organisms in ecotoxicological test systems assumes that genetic variability is related more or less directly to physiological and ecological variability. The validity of this assumption is likely to vary considerably. Also, it is important to recognize that the selection of a single genotype does not eliminate genotype × environment (e.g., toxicant) interactions; it simply ignores their existence. A significant genotype × environment interaction means that the phenotypic expression of a genotype depends on the environment in which it is found, or equivalently that the effect of the environment on the phenotype depends on the genotype (Maynard Smith, 1989). A significant interaction may appear as a change in the relative ranking of LC (or EC)$_{50}$ values among clones, depending on the specific toxicant tested (e.g., as in Baird *et al.*, 1991). Restricting testing to a single clone does not eliminate potential genotype × environment interactions (because the relative ranking of chemical toxicities will still depend upon which clone is selected for testing), but rather precludes their measurement (Forbes and Depledge, 1992a; 1993).

Studies with plants have shown that, under stressful conditions, small increases in factors such as water and nutrients can result in large growth responses. Thus the same level of environmental variation under stressful (e.g., toxicant-exposed) conditions can have a much larger influence on fitness traits. To complicate matters further, it is known that different genes may contribute to the same trait under different conditions (Hoffman and Parsons, 1991). This suggests that the inherent variability of single genetic clones will be dependent on culture medium and environmental conditions, including the chemical tested. In addition, mutations occurring in laboratory populations of organisms can rapidly lead to genetic divergence. Dobzhansky (1970) wrote, 'Inbred strains of experimental animals and plants are widely used in medical, physiological, and other research, in the belief that individuals of such strains are all alike genetically. This confidence is often misplaced because of high mutation frequencies.'

These discussions of standardization and variability need to take into consideration the ways in which experimental variance can be partitioned before protocols become incorporated into regulatory guidelines for environmental protection (Forbes, 1993). The first source of variability is the inherent biological variability of the test organisms. When considering the effect of a single toxicant on a population of organisms, it is the variance of the tolerance distribution itself. This biological component is the result of genotypic effects and genotype × environment interactions. The genotype × environment component is in turn influenced by the exact experimental conditions. Nevertheless, the essential point is that the variance of the tolerance distribution derives from the organisms and thus should be scrutinized closely and measured as accurately as possible in any evaluation of the effects of toxicants.

Such biological variability will play a central role in a population's ecological and evolutionary response to pollutant stress. It is this innate biological variability that is directly estimated by the probit regression (Finney, 1978; Forbes, 1993). If we restrict our attention solely to quantal concentration–response curves, the second and third sources of variability in a fitted model are referred to as 'lack of fit' (L) and 'pure' errors (E) respectively (Draper and Smith, 1981). Together they comprise the total residual variability in a concentration–response experiment ($T = L + E$). The above relationship holds for any statistical model. These two variance estimates then contain all the available information on the theoretical model's inability to explain reality and can be visualized as the residuals about the fitted line (in the case of a probit regression). It is E that is the classical experimental error variability and the proper focus of standardization efforts (Forbes, 1993).

What this means in practice is that attempts to standardize test systems and increase test reliability should focus on accurate and precise estimates of the statistical model. Both E and L should be made as small as possible. What we do not recommend are attempts to reduce the inherent *biological* variability of the system through the use of clones or highly inbred strains. This variability is the 'stuff of evolution' and should be studied carefully and included in toxicity estimates. All the usual caveats of good sampling design apply. Great pains should be taken to sample randomly from the population being characterized. Laboratory clones are fine if one is interested in determining the tolerance distribution of a laboratory clone. We must keep in mind, however, that the distribution parameters may be quite different from those of a natural population of the same species.

It cannot be taken for granted that standardization will eliminate or even substantially reduce variability in test results. Loomis (1975) observed up to threefold differences in the LD_{50} tested under carefully controlled laboratory conditions with the same batch lots of 10 different chemicals, the same detailed procedure and the same strain and sex of white rat. Furthermore, no increase in test variability was detected when each laboratory used its own standard test procedure. We should also mention, in connection with standardization, that efforts to achieve precise and accurate toxicity estimates have not proceeded equally in all directions. Whereas great efforts are expended to achieve a precise test result, such values are sometimes multiplied by an arbitrary application factor in assessing risk. Thus if the variability in the tolerance distribution is small relative to the application factor, there is little to be gained by further standardization. If, however the tolerance variability is large, then we need to analyse it more closely (Forbes and Depledge, 1992a).

4.6 Importance of power

The concept of statistical power is often overlooked in the design and interpretation of ecotoxicological tests. In the following paragraphs, we hope to demonstrate how 'powerful' test designs can improve the effectiveness of ecotoxicological test protocols and how knowing the power of a test can increase confidence in its result. These ideas can be illustrated by the following example. Suppose we are asked to determine whether the level of PCB in fish tissue is greater than the maximum acceptable level. Say that for reasons based on medical toxicological studies, the maximum tolerance level for the concentration of PCBs in fish tissue that will be consumed by humans has been set at 2 mg kg^{-1} fish tissue. The available data from a random sample of 30 fish give a mean of 2.1 and a standard deviation (SD) of 0.6 mg kg^{-1}. The question is, are the levels of PCBs in fish tissue greater than the maximum acceptable level or not?

To answer this question, the standard approach would probably be to compare the sampled PCB levels to the value of the environmental standard using a one-sided *t*-test for the difference between the sample population mean (the level of PCBs in fish tissue) and the single value of the legislated maximum. That is, one would define a null hypotheseis (H_O): [PCBs] $\leq$ 2.0 mg kg^{-1} and an alternative hypothesis (H_A): [PCBs] > 2.0 mg kg^{-1}. Then, choosing a significance level (in advance of the actual test and based on strong convention) of 0.05 (or 95%), one would conduct the test. Two results are possible.

First, our sampled PCB mean falls within a region of the *t* distribution such that the probability of our obtaining this particular result, if the null hypothesis is true, is less than 1 in 20 (0.05). In this case, there is no real problem and we can state with known confidence that the level of PCBs in fish tissue is greater than the environmental standard. We will only make the mistake of saying the value is greater than the standard, if in fact it is not, about 5% of the time. The almost exclusive focus on this type of error in modern hypothesis testing is due to the theory of statistical inference developed by the great biologist and statistician R.A. Fisher in the 1930s (Fisher, 1932). This philosophy has dominated applied statistical analyses since then. The approach is conservative in the sense that an alternative hypothesis must have strong supportive evidence in its favour before the null hypothesis is rejected. This is proper in most cases and prevents the development of theories on shaky ground.

The other possible result is that we fail to detect a significant difference between the population mean and the maximum acceptable PCB level. This is, in fact, the case with this particular test. Here the comparison of the sampled 2.1 mg kg^{-1} ($SD = 0.6$, $n = 30$) against the 2.0 standard, yields a P value of more than 0.18. So we cannot reject the hypothesis that true PCB concentration in fish tissue is 2.0 mg kg^{-1} or less at our predetermined level of significance. But does this mean that the level of PCB in fish tissue is safe? Can we say that the true population mean is probably 2.0 mg kg^{-1} or less? The answer is no. That conclusion would be strictly invalid. Yet the acceptance, either implicitly or explicitly, of an unfalsified null hypothesis is often seen in the literature. In order to answer the question of safe levels in light of our test results, we need to focus on the other possible error we can make, and then define and explore the concept of statistical power.

We contend that the error of accepting an unfalsified null hypothesis can have particularly serious consequences in ecotoxicological studies. Furthermore, we argue that a rational and conservative approach to ecotoxicological testing demands that we understand and quantify the probabilities associated with failing to find existing effects. This is in contrast to other, more traditional branches of science where it may be more appropriate to provide very strong evidence that a phenomenon exists before building on that knowledge.

In order to assess the safety of the PCB concentrations in our example above, we need to know something about type I and type II error rates as well as the concept of statistical power. Given the true state of nature and our statistical test result we either make the correct decision or an error (Table 4.2). Thus, in Table 4.2, the row probabilities sum to 1. A type I error is made when one concludes incorrectly that there is an effect when, in fact, there really is none. This is the type of error that scientists typically worry about. This is the error of concern in classic statistical inference when low significance criteria of 0.05 to 0.001 are set for the rejection of a null hypothesis. For example, this could happen if we concluded that the mean PCB concentration in fish was greater than the standard just because we were unlucky enough to obtain an improbable sample.

The probability of making a type I error is symbolized by α. For a given set of experimental conditions, α will be the limiting rate of false positive results as the test or experiment is repeated again and again. The strong convention exists to set this value at or below 0.05. The 95% significance level has achieved an almost sacred status in current approaches to statistical hypothesis testing. This convention is entirely arbitrary, however, and we note that there are cases for which one should consider relaxing this convention in the interest of minimizing type II error rates.

The type II error, which is typically under-rated in ecotoxicological studies, is made when a real effect is missed, that is, when we fail to reject a false null hypothesis. This is the error we would make if we failed to reject the hypothesis that the PCB level in fish tissue is less than or equal to 2 mg kg^{-1} when it actually is. In this example we did not reject H_0 because the P value was more than three times greater than our significance criterion ($P > 0.18$). If the true mean of PCB in fish tissue is greater than 2 mg kg^{-1} then we have just made a type II error. It is clear that in order to evaluate the safety of fish consumption we need to

Table 4.2 Decision table illustrating type I and type II errors

Truth	Reject H_0	Do not reject H_0
H_0 True	Type I error (α)	No Error ($1-\alpha$)
H_0 False	Power ($1-\beta$)	Type II error (β)

know more about the probability and consequences of making this particular mistake. Type II error is denoted by β. Given the falsity of the null hypothesis, β is the proportion of tests that will give a 'no effect' result as the test is repeated again and again under the same conditions.

One of the most important aspects of any statistical test is its power. The power of a test is its ability to detect an effect if there is one. It is equal to $1 - \beta$ (1 - the type II error rate). This is the feature of an ecotoxicological test that one should maximize so that the probability of missing a real effect is small. The power of a particular test is determined by several things.

1. The power of a test is proportional to the type I error rate, α, used for that test. This makes intuitive sense. If we test for the concentration of PCBs at the $\alpha = 0.01$ level we will be less likely to find a difference than if we test at the $\alpha = 0.05$ level, all other things being equal.
2. The power of a test also depends on the actual size of the effect. If there is no effect, the test will have no power to detect it. All things being equal, the larger the effect the greater the power of the test. Thus a critical aspect of power analysis is the determination of the effect sizes that one wishes to detect. In the example above any available medical toxicological data bearing on the chronic effects of ingestion of PCBs and fish consumption patterns of the consumers should be used in combination with measured fish tissue concentrations (preferably for a more extensive data set than $n = 30$) to predict intake levels and associated risks in human populations. This allows one to reasonably allocate resources for a sampling programme.
3. The third determinant of power is the reliability of the data (the variance, standard deviation, standard error, or whatever). No matter what the statistical test, the reliability is always proportional to the number of samples or replicates. The more fish we sample, the better our estimate of the mean [PCB] in the tissues of the actual fish population.

Thus power, which is a function of effect size, α and reliability, can and should be used to design sampling programmes. For example, if we take our PCB data where $n = 30$, $\alpha = 0.05$, $SD = 0.06$, and if the real sample mean is 2.1 mg kg^{-1} (which is our best estimate), then the power of the test is about one-quarter. This means that if the real PCB concentration is 2.1 we will fail to detect it (and thus make a type II error) in three out of every four replications of this particular sampling programme. It should now be obvious that there is an extremely good chance (75%) that the PCB concentration in our fish sample is above the maximum tolerance level even though we could not detect it. We can also ask the question, 'How many samples should we take in order to detect a real population value of 2.1 mg kg^{-1} 80% of the time?' The answer, again for $\alpha = 0.05$ and $SD = 0.06$, is that we would need to analyse 224 samples instead of 30. In this case it might be wise to relax α to say, 0.1 in order to increase the power of the test, especially if the cost of making a type I error is low.

Thus a well-thought-out response to the deceptively simple question, 'Are the fish over the tolerance limit with respect to tissue PCB concentrations?' requires analysis of the following factors.

1. What is the maximum effect size we are willing to tolerate? The higher it is the more easily we can detect it. The answer to this question depends on medical toxicological data and this should of course be taken into account when maximum tolerance levels are set.
2. What are the relative costs of type I versus type II errors? What is worse? Finding an effect that is not there or missing one that is?
3. Based on the answers to (1) and (2), what is the most appropriate sampling scheme or experimental design? Even if resource constraints prevent achieving the desired power due to sampling difficulties, inherent variability etc. we can at least attach probability values to alternative decisions. This can be a big help to decision makers.

Finally, we recommend that, before carrying out an ecotoxicological test or experiment, type II error rates should be estimated and the effects of the appropriate variables on the power of the test should be explored. An important goal in the design of any experiment should be to obtain maximum power per unit effort. Following the completion of a test or experiment, the biological ramifications of type II errors should be considered (what happens if the [PCB] is 2.1 or greater?). This will also help decision makers. If the null hypothesis cannot be rejected, *a posteriori* analyses should be performed. Questions such as, 'How does effect size influence power for this system?', should be addressed. Depending on the relative costs of type I and type II errors, it may be wise to vary α, the significance criterion, to a value greater than 0.05, such as 0.1. Perhaps the most important point is that **never** should it be concluded that the null hypothesis is true if a significant effect is not detected, unless it is done with full knowledge of the type II error rate.

4.7 Where is the eco?

The unfortunate fact remains that testing protocols purported to assess ecotoxicological effects of chemicals generally fall far short of their intended purpose. Giddings (1986) wrote,

> The ecologist approaches aquatic toxicology with a realization that organisms in nature are linked into complex systems that have structural and functional properties of their own, and that apart from the ecosystem context the effects of toxicants on natural populations cannot be completely understood. From this perspective, most standard toxicological methods are oversimplifications. The methods are not necessarily incorrect or inappropriate, but they are incomplete.

Giddings was neither the first nor last to express dissatisfaction with the lack of 'eco' in ecotoxicological testing strategies. Despite continuing protests of this kind and the knowledge that simplified laboratory test systems do not adequately represent chemical fate and effects in ecosystems, significant inertia remains to be overcome before improved testing protocols become incorporated into regulatory guidelines and management strategies.

Baker and Crapp (1974) stated that toxicity tests designed for the assessment of ecological effects have different requirements than those designed for ranking the relative toxicities of chemicals. The former must:

1. be designed for making predictions about a particular kind of community or habitat and should be based upon field studies in that community or habitat;
2. be capable of predicting mortalities among the key species of the community;
3. address the effects upon mortalities of such factors as seasonal or behavioural variations;
4. be based on a knowledge of the ecological consequences of the pollutant-induced mortalities.

Given the incompleteness of our current understanding, these criteria are much too ambitious to suggest for incorporation into routine testing requirements. Nevertheless, we believe they are reasonable requirements for research programmes in ecotoxicology. Strikingly, very few current programmes incorporate this minimum level of complexity. In Chapter 5 we illustrate an approach that, for aquatic sedimentary systems, functions at this presently neglected level of sophistication.

Implicit in the use of laboratory tests, as well as other tests designed to identify pollutant effects on single species, is the need to extrapolate from one or few species to the vast numbers of others making up a community. Moriarty (1983) recognized that, even if individuals of two species are affected by a pollutant to the same extent, the net effect of the pollutant on populations of the two species will be quite different because of differences in such things as use of resources and life history. This follows from the obvious but currently underappreciated and understudied fact that the components of complex systems interact. Landis (1986) argued that a major drawback of chronic laboratory toxicity tests (such as the *D. magna* 21-day test) is that they do not measure reproductive rates as a function of resource availability, which he contended was a crucial variable in the determination of the community-level impact of toxicants. Klerks and Levinton (1989) further complicated the picture by asserting that, in addition to interactions among trophic levels, the evolution of resistance to toxic substances can alter the potential transfer of toxic substances through the food chain and can result in effects on the community that are not predictable from standard toxicological studies. 'At worst, evolutionary response changes the predictions so completely that standard bioassays done on laboratory stocks are wholly inadequate, as they imply incorrect conclusions as to the fate of individual species and even the potential for transfer of toxic substances through the food web.'

Although our understanding of the impact of pollutants on populations, communities and ecosystems can perhaps be tested by efforts to extrapolate from the relatively simple to the more complex, we suggest that a better use of limited resources can be made by studying effects directly at the level of interest. For ecotoxicological research, this will be at the level of the population and above. Ecotoxicological research should be supported rather than driven by studies of the levels below. Biochemical and physiological investigations should be used to support and deepen our understanding of pollutant-induced changes in ecosystem processes or structural changes in communities. For example, biochemical studies can be used to evaluate the mechanisms behind the effects determined in community or ecosystem research. This position is exactly the reverse of the current *Zeitgeist* in ecotoxicological research. For example, there is a trend that is exemplified by the current extrapolation and validation fad to take lower level toxicological results and project them up to higher levels of complexity. Extrapolation beyond the end of a fitted line is a bad idea when employing regression models and an equally bad idea for ecotoxicological models. We must study the system of interest. If we can understand the system in its complexity we may be able to predict the effects of pollutants on it. If not, no amount of money spent on lower level biochemical or physiological investigations will improve the situation.

4.8 Tolerance as a quantitative phenotypic trait

4.8.1 Evolution of resistance to chemicals

Toxicologists recognized very early that variation among animals in their responses to (and in their tolerances of) drugs was as great as the variation in other measurable phenotypic characters. Detailed study of the tolerance distribution can reveal a great deal about changes in susceptibility or the evolution of resistance to toxicants. For example, study of the tolerance distribution as it changes with time can be a good way to examine the evolution of the stress response and can also aid in the study of the genetic changes that occur during toxicant exposure. The most detailed work to date has been in the area of the evolution of pesticide resistance by insect populations (e.g., Wood and Bishop, 1981). This work developed from the intense interest in the rates and mechanisms by which pesticide resistance evolved. As pointed out by Wood (1981) for insecticides, the evolution of resistance in response to exposure will typically involve physiological, biochemical and genetic components. In insect populations, the main mechanisms of resistance are (Maynard Smith, 1989):

1. detoxification (e.g., DDT is degraded by the enzyme DDT-dehydrochlorinase in genus *Musca* and some other insects);

2. alteration of the site of action (e.g., organophosphate insecticides act by inhibiting the enzyme acetylcholinesterase and many insects have evolved resistance through mutations altering the enzyme so that it is no longer inhibited);
3. reduced penetration into the organism (e.g., structural alterations in the cuticle of insects have been shown to decrease penetration of toxicants; the sequestering of heavy metals by metallothioneins within organisms may keep these substances away from sensitive metabolic machinery);
4. detection and behavioural avoidance;
5. increased excretion rates.

The earliest discussions of resistance debated whether or not natural populations could carry genes for resistance to chemicals they had never experienced before. Two distinct genetic bases for resistance, namely pre-adaptive (selection on genes already present in a population before chemical exposure) and post-adaptive resistance (arising from the 'unnatural' variation thought to be induced by the toxicant; Wood, 1981 and references therein), were proposed (summarized in Crow, 1957). The modern interpretation is that resistance arises from normal undirected mutation or natural variation already present in the population (Crow, 1957; Maynard Smith, 1989). In an elegant experiment, Luria and Delbrück (1943) determined that bacterial resistance to penicillin was based on alleles that had come about through mutation or were already present in the population. These results clearly demonstrated that mutations for resistance to non-mutagenic toxicants occur independently of exposure and are not induced by the toxicant (Futuyma, 1979). In another similar study of the evolution of DDT resistance in fruit flies, Bennett (1960) found that resistance in *Drosophila* was not induced by exposure to the chemical. Bennett bred siblings of the most resistant flies in each generation. These siblings had never been exposed to DDT. He found that after 15 generations the descendants were highly resistant to DDT. These results emphasize the fact that research into the evolution of toxicant resistance should take pains to sample, preserve and study this natural variability on which the response of natural populations is so dependent.

The story may be different for certain mutagenic toxins. Under certain conditions these chemicals may act to increase the development of resistance simply through increased overall mutation rates. Although not directed, the sheer increase in the mutation rate may improve the odds of resistance appearing. The probability of resistance occurring will of course depend on the potential mechanisms of resistance available to the organism and on the nature of the base or chromosome changes needed to bring it about. For example, the rate of evolution of parathion resistance in *Drosophila melanogaster* has been shown to increase in response to irradiation. The induced mutations occurred at a rate of 1.4×10^{-4} gametes (Kikkawa, 1964). This mutation rate change corresponded to roughly an order of magnitude increase over the background rate of approximately $1–2\times10^{-5}$ gametes or less. This suggests that one might expect faster evolution of resistance for toxicants having some degree of mutagenicity.

4.8.2 Analysing the shape of the tolerance distribution

If we consider the response of an organism to a toxicant to be a simple phenotypic character, then study of the shape of the population tolerance distribution can be used to indicate a response to selection. This conception of the tolerance distribution opens up for use by ectoxicologists the powerful set of theoretical and empirical tools developed by quantitative geneticists (e.g., Falconer, 1981; Via and Lande, 1985; 1987; Schluter, 1988; Anholt, 1991). Thus we can view individual tolerance as a standard quantitative trait. If the test population has been sampled well and is truly representative, the change in relative sensitivities of members of the population over time can be seen by comparing the distributions represented by the probit plots. The differences in shape reflect the underlying tolerance distribution in very specific ways (Wood, 1981). A great challenge in the future of ecotoxicology will be to attempt to link genotypic changes with distributional shifts in tolerance. In order to increase our understanding of pollutant effects on field populations, we need to be able to link genetic variation to the response to selection in the field. The knowledge and methods already obtained from population genetic studies on genetic variation, clines and natural selection (e.g., Koehn *et al.*, 1980; Falconer, 1981; Hilbish *et al.*, 1982; Hilbish and Koehn, 1985a; 1985b; Endler, 1986; Anholt, 1991) provide the necessary tools. This general approach, along with the attempt to provide a simple genetic mechanism for resistance, has been applied to insect populations. Wood (1981) suggested that one of the best ways to detect evolving resistance at an early stage is through close examination of the concentration–response curve.

When resistance is evolving through the spread of a gene or gene complex, which is still at low or intermediate frequency (10% or less), the EC or LD_{50} of a field-collected sample may still be quite similar to that of a susceptible strain (Wood, 1981). At this early stage it is much easier to detect changes in the shape of the concentration–response curve. Available diagnostic techniques for regression analysis make the detection of shape changes and departures from linearity quite simple and powerful. These changes will generally occur well before there are statistically significant changes in the EC_{50} itself (Forbes, 1993). An example of this type of response occurred in a field population of Danish houseflies (*Musca domestica*) exposed to the organophosphate pesticide, tetrachlorvinphos (Danish Pest Infestation Laboratory, 1970). As is generally the case when attempting to detect resistance in natural populations, the field-collected organisms from the insecticide-treated areas were compared with standard laboratory strains used as controls. The farm was sprayed with tetrachlorvinphos periodically during the summer of 1969 and the field population was sampled and compared with the laboratory strain. No resistance was detected at first but the flies showed a strong increase in resistance over the course of the summer. The field population showed changes characteristic of strong directional selection that were easily detected in the concentration–response curve. The increase in resistance of the field population was reflected in a very slight shift of the LC_{50} to the right, but a strong flattening of the curve that increased with time. This change in the probit plot reflects a skewing of the tolerance distribution to the higher toxicant concentrations in the way that would be expected with strong directional selection. Interestingly, the population appeared to stabilize after about six months of exposure to the insecticide even though it was still very heterogeneous in response. By October roughly 50% of the flies were resistant to tetrachlorvinphos concentrations that would have killed 97% of the flies in May.

To illustrate this principle further, we can generate a graph of predicted changes in the concentration–response curves with a simulation of directional selection in a population of the grass shrimp *Paleomonetes pugio* exposed to the pyrethroid insecticide fenvalerate. We first construct a control tolerance distribution for the grass shrimp that is normally distributed about a mean fenvalerate concentration of 4.00 μg l^{-1} ($\sigma_T = 1.00$ μg l^{-1}). We then characterize the control population by taking a random sample of 1000 individuals before fenvalerate exposure. The resulting histogram is depicted in Figure 4.6a. This sample tolerance distribution closely approximates the normal curve with a mean of 3.98 mg l^{-1} and a standard deviation of 1.001 mg l^{-1}. If we plot the probits against fenvalerate concentration, we find that it very nearly follows a straight line, except for the points near the extremes of the distribution, which tend to fluctuate just a bit (Figure 4.6b). This is precisely what we would have expected for a population whose tolerance distribution closely approximated normality. If the population were exposed to 4 μg l^{-1} fenvalerate over several generations, this would exert very strong directional selection for increased resistance (e.g., Falconer, 1981; Futuyma, 1979). Selection for resistance would be revealed as some form of directional shift in the tolerance distribution toward the higher concentrations of fenvalerate. The shift could take the form of a displacement or skewness (e.g., Wood, 1981), and could be most easily studied as a change in shape and/or location of the probit plot for the succeeding generations of shrimp.

We can explore the situation further by using simple graphical methods for analysing distributional characteristics that take advantage of the fundamental nature of the probit plot. To generalize, we note that the probit plot is simply a specific example of a class of plots called theoretical quantile–quantile or probability plots (e.g., Chambers *et al.*, 1983). This allows inferences about the underlying tolerance distribution to be made based on a simple graphical analysis of the plot. The probit plot itself is simply a graph of the corresponding percentage points of the normal distribution (plus five to give non-negative

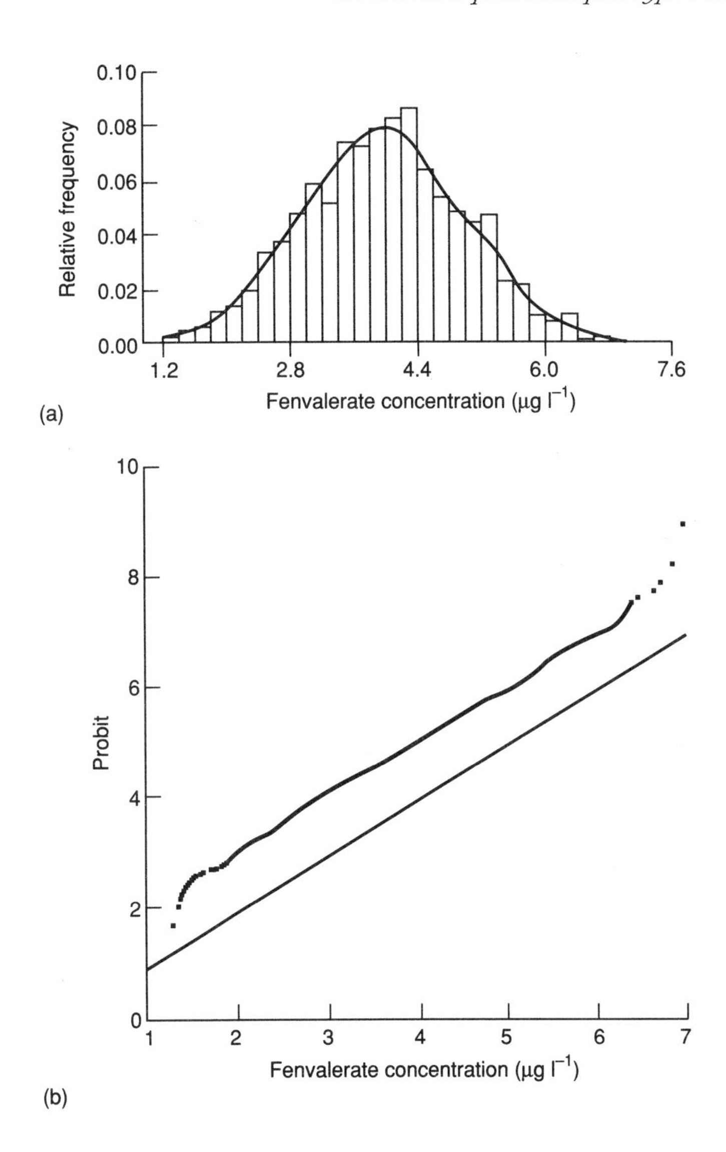

Figure 4.6 (a)Tolerance distribution for the response of *Paleomonetes pugio* to the insecticide fenvalerate. Where $\mu_T = 3.98\ \mu g\ l^{-1}$, median $= 3.98\ \mu g\ l^{-1}$, $\sigma_T = 1.00\ \mu g\ l^{-1}$ and the total population (sample) size $(n) = 1000$. This population was drawn randomly from a population with a mean of 4 and standard deviation of 1. The continuous smooth of the data was estimated using a nonparametric density estimator (Silverman, 1986). (b) Probit plot. (■) = Probits for each of the individual members of the population shown in (a). The solid line is $y = x$. See text for further details.

probits) against the toxicant concentration (Bliss, 1935; Finney, 1978; Forbes, 1993). It is because of the feature that toxicant concentrations are plotted against their 'expected values' based on the normal distribution, that the points tend to follow a straight line if normality of the tolerance distribution is closely approximated. Thus the straightness of the plotted points in Figure 4.6b clearly suggests a normal tolerance distribution. In addition to a quick estimate of the median (or mean) usually made by analysts, we can compare the plotted points to the line $y = x$ (Figure 4.6b). A plot parallel to $y = x$ indicates a standard deviation for the tolerance distribution of approximately 1.0 whereas clockwise rotation suggests a standard deviation greater than 1.0 and counterclockwise rotation indicates a standard deviation less than 1.0 (Chambers *et al.*, 1983). Thus a quick glance at Figure 4.6b suggests an approximately normal, underlying tolerance distribution whose standard deviation is close to 1.0 and whose mean (and median) is very close to 4.0, as is in fact the case.

If the population is then subjected to directional selection for increased toxicant resistance, the shape of the probit plots of subsequent generations can change in several interesting ways (Wood, 1981). Figure 4.7a is a simulated tolerance distribution depicting the evolution of resistance after the previous population has responded to directional selection. There is now a small fraction of highly resistant individuals, skewing the new population to the right (cf. Figures 4.6a and 4.7a). The probit plot for these data is shown in Figure 4.7b. An OLS regression through these points indicates a decrease in the estimated LC_{50} (3.6 vs 4.0 $\mu g\ l^{-1}$). A second estimate based on direct interpolation from the probit scatterplot also shows a decrease in median effect concentration down to approximately 3.3 $\mu g\ l^{-1}$. Nevertheless, the resistance-skewed underlying tolerance distribution is shown quite clearly with the break in, and flattening of, the probit plot. This is the same effect observed for the actual development of resistance to trichlorvinphos by Danish houseflies (Danish Pest Infestation Laboratory, 1970). Quantitative genetic analysis of this type of resistance in natural populations has generally indicated that resistance is due to one or a few genes (Maynard Smith, 1989).

In this example, the density of sampling used to construct the probit lines in Figures 4.6b and 4.7b is quite high (1000 points) and this fact allowed us to assume with very little error that these distributions represent the true ones. Sampling at this density makes departures from linearity quite easy to see. In a real experiment we would have to use many fewer points to construct the plot for estimating the tolerance distribution, and this would be expected to make any nonlinearities much more difficult to detect. Is it realistic to expect to be able to detect this degree of change using a practical number of samples? In general it should be possible if we take the following approach.

In a standard acute toxicity test a sound design would require focusing our sampling in two main regions of the tolerance distribution. We would probably want to sample intensively near the expected LC_{50} as well as near the 15% and 85% response or effect levels. This design represents a compromise strategy

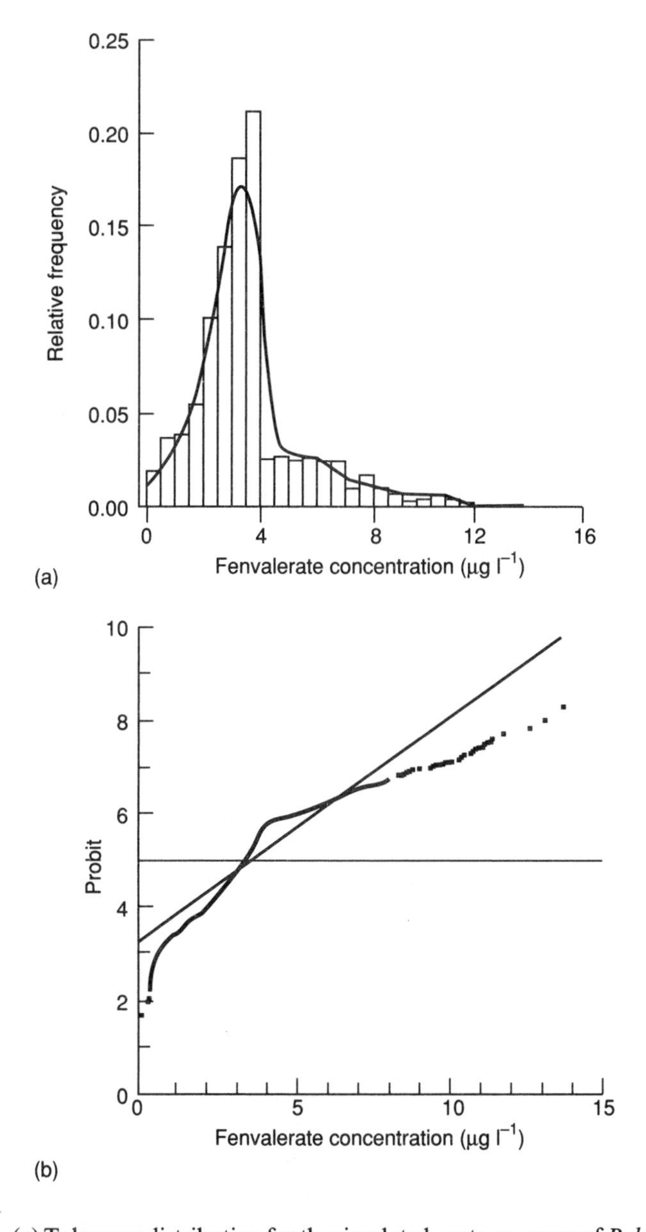

Figure 4.7 (a) Tolerance distribution for the simulated acute response of *Paleomenetes pugio* to the insecticide fenvalerate after some degree of resistance has been allowed to develop. The 'total population size' is again equal to 1000. The continuous smooth curve of the data was estimated using a nonparametric density estimator (Silverman, 1986). (b) Probit plot. (■) = Probits for each of the individual members of the population shown in Figure 4.6a. The horizontal solid line is probit = 5.0. The angled solid line is the ordinary least squares regression fit to the probit data. See text for a detailed discussion.

due to the tradeoffs involved in obtaining a reliable estimate of the LC (or EC_{50}) (Finney, 1978; Forbes, 1993). However, in the present instance, we are most interested in determining whether or not there has been a shape change in the underlying distribution. Therefore we will sample intensively within approximately one-half of a standard deviation to either side of our best estimate of the EC_{50} (EC_{50} around 4.0 μg l^{-1}, Figure 4.6a and b). We decide on three replicates at each of 10 treatment concentrations centred about 4.0 μg l^{-1}. We further allot 10 *P. pugio* to each replicate, for a total of 300 shrimp in all. We simulate this experiment, again sampling from a binomial distribution using the known proportion of shrimp killed at each concentration (based on the distribution in Figures 4.7a and b).

The question we ask is the following. Is it possible to detect the resistance that has developed using this experimental design? We see from Figure 4.8 that even with the considerable variability at this level of resolution we can quite clearly see the result of directional selection. The open circles connected by the smooth curve are the probits calculated from the fraction of the population to the left of the fenvalerate concentration for the distribution shown in Figure 4.7a. The filled circles are replicate samples from the binomial distribution centred along the smooth curve. If we ignore the two solid lines for a moment and focus only on the filled circles, it looks like there might be a levelling-off or flattening-out of the scatterplot.

But how do we know this is not just a chance effect? The key feature to note in this figure is the slightly jagged line that more or less closely follows the smooth curve. The approach we have taken here which generated the jagged line was the application of a locally weighted scatterplot smoothing algorithm or **LOWESS** (Cleveland, 1979; Chambers *et al.*, 1983; Cleveland and McGill, 1984). This line is a robust estimate of the data points and is actually a form of nonparametric regression (Efron and Tibshirani, 1991). Such an approach is ideal for this kind of data because, unlike linear or polynomial regression, no form of dependence of y on x is assumed.

The smooth was estimated as follows. Like the regression procedure for calculating the probit regression line, LOWESS also employs weighted least squares. The smoothing procedure produces estimates of the centre of the distribution of replicate probit values at each concentration by calculating a series of local, weighted linear regressions. The weighting means that the points corresponding to nearby concentrations have the most influence, thus making the estimate 'local'. In an analogous fashion, points receive weights in the y direction as well. These estimates are then connected to produce the more angular continuous line shown in Figure 4.8. The estimates are both locally weighted and robust. This means that outliners and data points distant from the midpoints of the local regressions are much less influential, allowing the estimates more freedom to follow the local trend indicated by the data.

Figure 4.8 demonstrates that this approach provides an extremely accurate estimate of the underlying functional relationship despite the very great variability

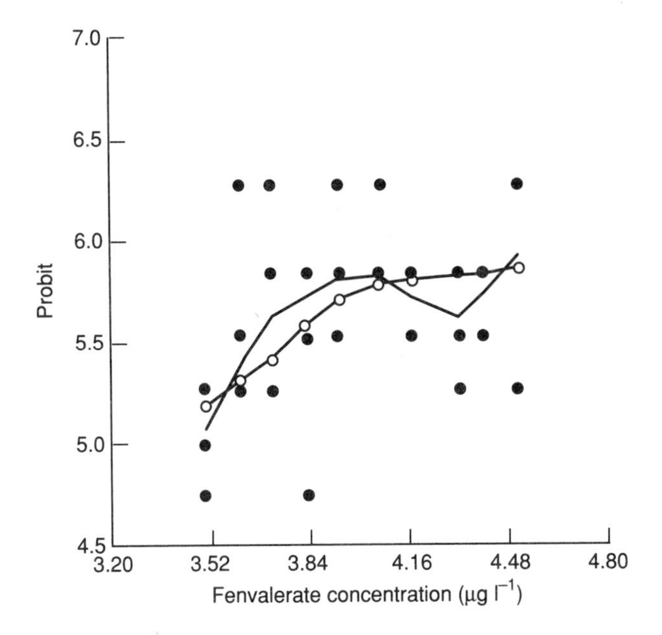

Figure 4.8 Probit plot based on random samples taken at 10 different treatment concentrations from the tolerance distribution shown in Figure 4.7a. Four replicates based on 10 shrimp each are plotted as filled circles. The smooth curve connected by the open circles is the distribution from which the samples were taken. The more jagged line is a LOWESS smooth. See text for details.

inherent in the sampling regimen and experimental design (Figure 4.8). Our hypothesis is that the data points will fall on a curved line, tending to flatten out at the higher fenvalerate concentrations. Trying to make a visual and subjective assessment of the curvature of the data points would prove inconclusive. This is equivalent to attempting to determine a treatment effect by examining the raw data. Inspection of the LOWESS smooth shows that, indeed, there is strong evidence that directional selection has occurred. It is important to use a technique such as LOWESS in determining the functional relationship in an effort to detect selection. This is because we are unable to postulate a functional relationship, and we need to be able to cut through the large amount of variability that is often present in real data.

4.9 Summary and conclusions

There is considerable room for improvement in current ecotoxicological testing schemes. The foundation for toxicology and ecotoxicology is provided by the dose (or concentration)–response relationship. Despite a long history of development, the design, analysis and interpretation of concentration–

response relationships (particularly for quantal data) are frequently performed poorly and misunderstood. Acute toxicity tests of single species are widely used to assess ecotoxicological effects. There is doubt whether acute toxicity can be used to predict chronic effects on the same species with a useful degree of certainty. Great efforts are being expended to improve the standardization of international testing protocols. More effort needs to be devoted towards improving the statistical design and analysis of tests and less should be devoted towards removing inherent biological variability from test populations. Ring tests can be useful for ensuring that different laboratories maintain the same high standards of scientific quality. However, the failure to distinguish between the development of protocols for testing laboratories and those for testing organisms can lead to the implementation of inappropriate guidelines. The factors contributing to test variability can be partitioned into:

1. inherent biological variability among organisms;
2. lack of fit to the model;
3. experimental error.

It is the latter two sources of error, but not the first, that should be the focus of standardization efforts. The inherent biological variability can and should be analysed in order to investigate the potential of populations to evolve resistance to chemicals and to study the underlying mechanisms contributing to resistance.

The error associated with failing to reject a false null hypothesis (type II), can have particularly serious consequences in ecotoxicological studies, yet this type of error is largely neglected. Maximization of statistical power (the ability to detect an effect if there is one) per unit effort should be a priority in designing ecotoxicological experiments. This will require that the desired effect size, type I error rate and variability of the data be taken into consideration. It is never valid to conclude that the null hypothesis is true if a significant effect is not detected, unless it is done with full knowledge of the type II error rate.

5 *Integrated ecotoxicology: linking fate and effect within a biological hierarchy*

Finally, no one who considers the facts given in this chapter ... on the vast number of worms which live within a moderate extent of ground ... on the weight of all the castings ejected within a known time on a measured space – will hereafter, as I believe, doubt that worms play an important part in nature.

Charles Darwin, 1896

5.1 A fundamental problem

Ecotoxicology is an infant science. Its continued growth and development requires that strenuous efforts be made to extend the manner in which it deals with current research problems. Advances in our understanding of the complex environmental problems that we face today demand that the fate and effects of pollutants in ecosystems be approached as an integrated whole. These two aspects of the same problem cannot be separated. In our view, ecotoxicology has failed to develop sufficiently beyond its roots in toxicology. While correctly focusing a great deal of attention on the fate of pollutants in the environment and the toxic effects of chemicals at the level of the individual organism and below, for example, on the mechanisms of toxicant action at the cellular and subcellular levels, ecotoxicologists have generally been slow to develop investigations of the complex linkages between the two. As a result, questions dealing with processes occurring at levels above the individual have continued to prove extremely difficult to answer.

With the current state of the art we find ourselves in the unfortunate situation in which the scientific basis of environmental regulation and management of pollutants is almost entirely based on very simplistic laboratory tests. This situation exists despite the fact that we are often aware that the results of these tests marginally reflect the complex and interactive processes occurring in nature. We are forced into relying on what we know to be inadequate regulatory tools because of our fundamental ignorance of the way in which toxicants affect the dynamics of natural ecosystems.

One approach to this problem has been the advocacy of research into the 'validation' or 'extrapolation' of the results of standardized toxicity tests to higher level processes actually occurring in nature (e.g., Stephan *et al.*, 1985; Kooijman, 1987; Van Straalen and Denneman, 1989; Cairns, 1988; 1990; Aldenberg and Slob, 1991; Wagner and Løkke, 1991; Forbes and Forbes, 1993). This school of thought focuses on demonstrating the validity (or lack thereof) of the current toxicity evaluation methods at higher levels of biological order. Presumably, any lack of correspondence between the test results and pollutant effects on ecosystems or natural populations can be repaired by developing new test systems (perhaps by incorporating more complexity) or tweaking old protocols. To the extent that it focuses the thought of scientists and regulators on more complex systems, this is a very good thing. In some cases one may then be able to rely on monitoring programmes to ensure that the agreed-upon levels of environmental damage have not been exceeded (Cairns, 1990).

It is clear that in order to validate or extrapolate results from the simplistic tests used for regulatory purposes to higher levels of complexity, we must be able to define what constitutes a damaged or undesirable ecosystem, community, or assemblage. This definition will always contain an irreducible component of subjectivity (Chapter 3). It is impossible to find a biotic system on Earth today which displays no evidence of human activity. Thus the critical questions must concern how much pollution we will accept rather than how much we can pollute without effect. An additional and perhaps obvious point is that the quality of this inherent subjective component will be directly proportional to our fundamental knowledge of the effects of pollutants in nature. To the extent that we understand the effects of pollutants on ecosystems, it will be possible to choose the kind of environment in which we wish to live.

5.2 The approach

The purpose of this chapter is to focus on approaches for linking biological effect with pollutant fate at the level of the organism and higher. If predictive or quasi-predictive models can be developed for higher level biological processes, we will be in a better position to develop criteria for environmental management and stewardship.

We will illustrate the direction in which we feel research should proceed through the use of illustrative examples from the field in which we are most familiar, aquatic ecology. The goal is to set up a tentative, quantitative framework for investigating the complex interactions involved in linking fate and effect. We begin by examining pollutant diagenesis in marine sediments. By diagenesis we refer to the sum total of physical, chemical and biological processes affecting the fate of a pollutant after it first arrives at the sediment–water interface (Berner, 1980). This approach is well suited for our need to integrate all of the above mentioned components at the ecosystem level. Examination of pollutant fate will primarily be through the use of so-called diffusion–reaction–advection models which have been developed by marine geochemists and chemical oceanographers during the past 20 years or so. In addition, we will begin to link the effects of pollutants on the organisms inhabiting sediments with pollutant distribution and/or degradation. This will be accomplished by the incorporation of specific terms in the model equations that can be used to couple fate and effect. Marine sediments are an ideal environment in which to study the interactions between the fate and effects of toxicants because of the extremely tight geochemical coupling between infaunal inhabitants and the sedimentary environment.

Approximately 70% of the Earth's surface is covered with marine sediment. Most pollutants, unless rapidly degraded, eventually find their way to the soft-bottom benthic environment (e.g., Rice and Whitlow, 1985a). The final fate of these pollutants will be determined by their effects on, and interactions with, infaunal organisms and biogenic structures. A great deal of research has focused on the relationships between concentrations of common pollutants in sediment particulates and pore waters and their effects on benthic organisms. Yet we are far from a predictive science of chemical toxicity in marine sediments. This is because past studies have tended to deal primarily with the fate *or* effects of anthropogenic contaminants and have been directed at levels below the population and/or ecosystem (e.g., Swartz, 1987; National Oceanic and Atmospheric Administration, 1989). This failure to explicitly examine the interactions between the two aspects of the same problem at the population or ecosystem level has been at the core of our lack of understanding of pollutant behaviour and effects in natural sedimentary systems.

Importance of benthic organisms in determining the fate and effect of pollutants in marine sediments The effects of invertebrates on the physicochemical properties of sediments and terrestrial soils can be traced back at least 100 years to the work of Charles Darwin (e.g., Darwin, 1896). It is therefore surprising that there appears to be a general disregard for the effects of resident metazoans on pollutant fate and degradation in the ecotoxicological literature. With very few exceptions the approach to pollutant chemistry in soils and sediments has been one of shaker flasks and sediment slurries. Even using these simplified approaches, effect and fate studies have tended to proceed in parallel with relatively few interactions and have failed to consider how the chemical milieu, in which the benthos live and exchange matter and energy, develops the physicochemical characteristics that must ultimately determine pollutant behaviour (Rice and Whitlow, 1985b).

Because the chemical and physical evolution of a deposit is profoundly influenced by the biogenic structures present, the organisms inhabiting the sediment must exert some degree of indirect control on the effect of pollutants on themselves (Aller, 1988; Rice and Whitlow, 1985b). The exploration of these linkages in some detail is the principal focus of the following examples. The interactions among the biota, the sediment and pollutants are influenced by several important effects that organisms have on the sedimentary environment. An obvious example is the construction of burrows and the resulting physical alteration of sediments (Rhoads and Boyer, 1982). Burrow structures have been shown to increase the total anoxic–oxic boundary area as well as the total surface area for diffusive exchange with the overlying water column. Thus biogenic structures profoundly change the relative dominance and distribution of important oxidation–reduction reactions, such as NH_4^+ and HS^- oxidation, and increase overall biogeochemical heterogeneity. These factors will be most important in anoxic coastal sediments where the impact from land-based pollutant loading is greatest. Perhaps most important are organism-related changes in:

1. sediment bulk composition and physical properties;
2. pore water solute profiles;
3. reaction rate distributions within the sediment;
4. fluxes across the sediment–water interface (Aller, 1988; Rhoads and Boyer, 1982).

These factors are necessarily tightly coupled to changes in microorganism activity and abundance. For example, it is well known that the zones around many biogenic structures are often the sites of greatly enhanced microbial activity and increased meiofaunal and microbial population sizes (Aller and Aller, 1986; Kristensen *et al.*, 1985; Aller and Yingst, 1978; Hylleberg, 1975). The central importance of the benthos in the control of the physical and chemical characteristics of aquatic sediments is illustrated diagrammatically in Figure 5.1.

If we follow the approach previously taken by oceanographers and geochemists, we can model these biotic influences as consisting of the summation of effects due to individual animals on diffusion geometry (in the case of solutes) or bulk sediment reworking and bioturbation rates (in the case of particulates) (e.g., Aller, 1980; 1982b; 1988; Guinasso and Schink, 1975; Cochran and Krishnaswami, 1980). In the examples below we will view the biodegradation of a pollutant as dependent on its own reaction kinetics *as well as* on the biological and physical mixing and alteration of the sedimentary matrix. As a model pollutant, we have chosen to investigate polycyclic aromatic hydrocarbons (PAH) from the organic family of compounds because they are both common and persistent, and many are known carcinogens. In addition, there often seems to be a perverse tendency to consider even these very persistent contaminants to be 'out of the ecosystem' once they are incorporated into

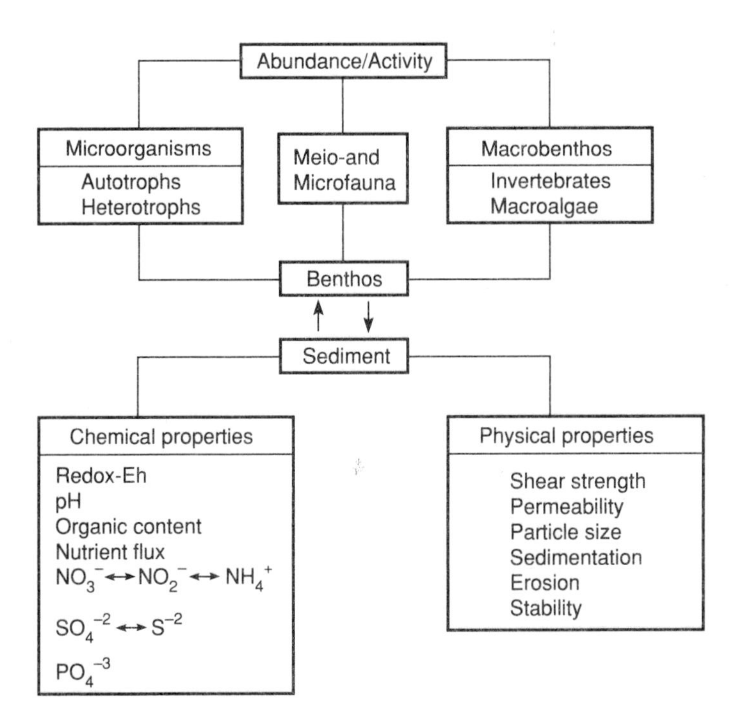

Figure 5.1 The sediment and benthos as an integrated and dynamically interacting system. Adapted from Meadows and Tufail (1986).

sediments. The fallacy of this viewpoint is well illustrated by work on the phosphorous budget of Lake Erie, USA. In 1975 the United States Army Corps of Engineers published a report based on a detailed investigation of the phosphorous budget for the lake, which indicated that the flux of phosphorus from the sediments to the lake water was greater than the total external loading due to surface runoff and aerial deposition (cited in Matisoff, 1982). This example highlights the fact that our ability to predict the future behaviour of the vast reservoir of pollutants currently incorporated in sediments will be critically dependent on our understanding of pollutant–animal–sediment relationships. With this in mind, a secondary goal of this chapter is to demonstrate in detail the kinds of complex interactions that combine to determine the fate and effects of pollutants once they leave the water column and enter the sedimentary environment. Not only is the soft-bottom benthic environment a valid, living system in its own right, but exchange between sedimentary and water column systems is time-dependent, dynamic and strongly influenced by the benthos.

Our example concerns the degree to which infaunal animals biologically mix the sediment and how this in turn can be related to pollutant fate and effect. We have chosen an example in which the pollutant is expected to be strongly associated with sedimentary particulates. This was partly because the equations for the relationships that allow us to begin to couple pollutant fate and effect are in general simpler than those describing solute behaviour. However, it is important to note that the modelling process applied below could equally well be carried out with pollutants in solution. In fact, in some ways solutes are easier to deal with because these models can incorporate variables that quantify animal density and size and hence are capable of allowing smooth coupling between the individual and population levels (e.g., Aller, 1988). These more complex models dealing with solute concentrations have the added advantage of allowing predictions of biodegradation rates in and around individual animal burrows and fluxes at the sediment–water interface, providing a solid basis for integrating studies in both the benthic and pelagic realms. Note also that we have downplayed the role of physical mixing for simplicity of presentation. A comprehensive treatment of the fate and effects of particle-bound organic pollutants in nearshore environments would need to consider physical factors in much more detail than we have treated them here.

A central challenge in the future will involve working out the details of the feedback loops between fate and effect for different classes of pollutants. Our goal is to use the modelling approach briefly outlined below to isolate the critical unknown parameters and specific gaps in our knowledge that need to be filled in order to meet this challenge.

5.3 Diagenesis of polyaromatic hydrocarbons in marine sediments

Polycyclic aromatic hydrocarbons (PAH) constitute a broad class of environmentally persistent organic compounds that are now ubiquitous in both aquatic and terrestrial ecosystems (Shiaris, 1989). Many are known carcinogens in mammals (Neff, 1979; 1985). The environmental impacts of the many PAH that have not yet been demonstrated to be carcinogenic are still largely unknown (Neff, 1979). PAH are released into the environment in large part due to human activities such as the combustion of wood and fossil fuels (Rand and Petrocelli, 1985). Structurally, these compounds consist of two or more fused benzene or aromatic rings; a bonding geometry that results in an overall planar aspect. The ultimate fused-ring form is graphite, the allotropic form familiar to most people as pencil 'lead' (Neff, 1985). As a general rule, the lower molecular weight PAH tend to be more volatile, soluble and relatively mobile in the environment. Even so, most are extremely insoluble and tend to be associated with particulate matter; especially sedimentary organics (e.g., National Oceanic and Atmospheric Administration, 1989; Swartz *et al.*, 1990). Vapour pressure, aqueous solubility, susceptibility to oxidation and reduction all tend to decrease with increasing molecular weight. Vapour pressure and solubility decrease logarithmically (Neff, 1979). These physical characteristics mean that the various classes of PAH can differ substantially in their effects on biological systems and behaviour in the environment. For example, the lower molecular weight compounds, with two to three aromatic rings, such as the naphthalenes, anthracenes and fluorenes tend to be the most acutely toxic, whereas the larger PAH with four to seven rings, such as benzopyrenes, chrysene and coronene, are not as acutely toxic but tend to be more carcinogenic (Neff, 1979; Futoma *et al.*, 1981).

Many environmental and biological factors are known to influence PAH biodegradation rates in sedimentary environments. These include oxygen concentration, temperature, salinity, the local hydrodynamic regime, PAH concentration, organic matter concentration, nutrient concentration and the degree to which microbes are grazed by detritivores (Hambrick *et al.*, 1980). All of these factors can be influenced to some degree by infaunal benthos and many of the most important, including microbial activity and oxygen concentration, are literally dominated by effects of benthic animals (e.g., Aller, 1988). In addition, microcosm experiments have demonstrated that including benthos in sediments can significantly enhance degradation rates; though the exact mechanisms remain hypothetical (Gardner *et al.*, 1979; Bauer *et al.*, 1988). For example, rates of microbial respiration, glucose assimilation and degradation of anthracene were enhanced in marine microcosms containing the polychaete *Capitella capitata* (Bauer *et al.*, 1988).

With the exception of the previously mentioned two studies, measurements of PAH degradation in sediments have generally been performed under extremely artificial conditions. Typically these studies measure degradation rates in dilute, afaunal sediment slurries (e.g., Shiaris, 1989). Under dilute, aerobic slurry conditions degradation rates can be several orders of magnitude greater than those measured in microcosm experiments. Gardner *et al.* (1979) measured degradation constants on the order of 10^{-8} s^{-1} for anthracene, fluoranthene and several benzopyrenes in the presence of *Capitella capitata*. In contrast, dilute slurry-based measurements give decay rates of 10^{-5} s^{-1} or so for the same compounds (Shiaris, 1989). In terms of their ecotoxicological usefulness, afaunal slurry results are equivocal and must be interpreted with caution. Nevertheless, some overall trends are apparent. PAH show an almost universal resistance to biodegradation and become increasingly refractory under reducing conditions (Hambrick *et al.*, 1980; DeLaune *et al.*, 1981; Shiaris, 1989). Coupled with a strong tendency to associate with organic matter, lack of degradability under anaerobic conditions enhances the problematic nature of these compounds. They will break down most slowly in precisely the environments where they will be added in greatest amounts. Later in the chapter we will employ geochemical modelling to suggest that PAH may also tend to accumulate in those benthic communities containing the most active sediment reworkers; the so-called *K*-selected or late successional communities that are often associated with otherwise low disturbance levels (e.g., Rhoads *et al.*, 1978).

5.3.1 Sources of PAH to the benthic environment

Primary formation of PAH can occur by a variety of processes. These include indirect and possibly direct biosynthesis, diagenesis of organic matter in sediments to produce fossil fuels, and incomplete combustion of fossil fuels (Neff, 1979). Indirect biosynthesis occurs when polycyclic compounds, which are probably derived mostly from plant pigments, are exposed to the reducing environments of anoxic sediments. Thus when PAH derived from fossil fuel combustion are deposited and subducted into the anoxic layers of recent marine sediments, they are completing a carbon cycle initiated hundreds of millions of years ago.

In 1979 an estimated 230 000 metric tonnes of total PAH were thought to enter the aquatic environment annually (Neff, 1979). These compounds can reach sediments through two major avenues. In environments distant from the source of combustion, PAH associated with particulates are generally transported by the wind and deposited through rainfall and dry fallout (Gschwend and Hites, 1981). Those that are deposited on the sea surface then settle and become incorporated into the sediment; a process that probably began in earnest sometime close to the turn of this century (Gschwend and Hites, 1981). For example, aerial transport is probably responsible for the present day delivery of individual PAH to remote sites in northeastern Maine, USA at fluxes of 1–3 ng cm^{-2} per year (Gschwend and Hites, 1981). For sedimentary environments closer to the pollutant source, contaminant flux is probably dominated by erosion and surface runoff of sorbed PAH (Windsor and Hites, 1979; Gschwend and Hites, 1981). Flux measurements for heavily polluted coastal environments such as Boston harbour and Buzzards Bay, Massachusetts, USA are in the neighbourhood of 1–140 ng cm^{-2} per year for most individual PAH (Gschwend and Hites, 1981).

Once PAH reach the aquatic environment, they can be degraded or removed by evaporation, photo-oxidation, metabolic degradation by aquatic microorganisms, fungi or animals (Neff, 1979; Farrington, 1980). The lower molecular weight compounds such as the naphthalenes and phenanthenes tend to be removed from the water column primarily by evaporation, microbial oxidation and sedimentation. The higher molecular weight compounds such as the benzopyrenes are probably removed chiefly through sedimentation and, to some degree, by photo-oxidation (Neff, 1979). Neff has estimated that approximately 50% of the higher molecular weight PAH entering the water column eventually end up in the sediment.

Because of their extremely low solubilities, any PAH entering the marine environment in solute form will be adsorbed rapidly onto particle surfaces where they tend to be strongly associated with particulate organic matter (Neff, 1979; DiToro *et al.*, 1989; Swartz *et al.*, 1990). Thus, *a priori* one expects them to accumulate in areas of relatively high particulate organic matter (POM) deposition. Low degradation rates often lead to high degrees of accumulation in sediments. This leads to the often observed situation of low water column, but high sediment and tissue contaminant concentrations. Aquatic organisms often contain tissue concentrations orders of magnitude higher than aqueous concentrations, but similar to or slightly less than those of the bottom sediments they inhabit (Neff, 1979; Neff and Anderson, 1981).

The observation that PAH tissue concentrations are often slightly lower than the concentrations found in sediments may be due to the kinetic balance between metabolism of the compounds and their rate of supply or uptake. Many organisms may be able to maintain steady-state concentrations slightly below those expected based purely on equilibrium partitioning into tissue. Testing the validity of this hypothesis and evaluating the costs incurred could be a fruitful area of future research. The results of such experiments could be directly related to theories dealing with the partitioning of energy between growth, reproduction and the metabolism of toxicants (e.g., Sibly and Calow, 1986). We suggest this as an example of the type of approach that is needed in order to place the mechanistic basis of chronic or sublethal pollutant effects on a more solid theoretical foundation (Chapter 6).

Owing to the extremely low solubilities of most PAH in seawater, it is probably appropriate to view sediments as a vast contaminant reservoir releasing minute amounts of material into the dissolved phase as PAH is taken up by the animals or degraded in solution. In heavily polluted sediments, this must place a long-term burden on the metabolic systems of the inhabitants. As far as we have been able to determine, the degree to which this can contribute to chronic, sublethal effects at the organism, population and community levels in natural systems is unknown.

5.3.2 Some initial simplifying assumptions

Because of the intimate association between sedimentary organic matter and PAH, we will begin our analysis by adopting the approach pioneered by Rice and Rhoads (1989) to model the early diagenesis of reactive (biodegradable) organic matter in coastal sediments. Our first example consists of a detailed look at the effect of bioturbating benthos on the fate of particle-sorbed PAH. Initially, we will assume an exponential decline in the concentration of the particle-sorbed contaminant with depth due to microbial degradation and simple first-order kinetics. For simplicity, we will also view benthic bioturbation (and physical mixing) as homogeneous and constant as a function of depth. This is the most straightforward formulation and is often quite successful in modelling particle reworking due to animal activity (Aller, 1982b; Rice and Rhoads, 1989). Because of their association with organic matter, PAH will be constantly supplied to the sea bottom along with the depositional rain of particulate organic matter where they will tend to accumulate. We will also make the simplifying assumption of a steady-state depositional flux. Later, we will briefly explore some of the consequences of changing depositional flux as a function of time.

Once a particle-associated PAH reaches the seafloor, additional complications can arise. The pollutant can undergo resuspension and transport, incorporation into the deposit, redistribution due to the activities of the benthos, and chemical reactions within the sediment itself (Berner, 1980; Aller, 1982b). To lay the groundwork, we first examine the simple case where microbial degradation occurs in the complete absence of reworking by benthic animals. Even this simple approach adds a two-dimensional complexity wholly absent from most slurry-based laboratory studies. It should be remembered that, although we discuss PAH contamination as a specific example, the following discussion is equally valid for all particulate or particle-bound degradable pollutants released into the aquatic environment. We are using marine sediments as a model system, but the general concerns regarding relationships among physicochemical and biological variables must be dealt with in any integrated ecotoxicological study. We begin by investigating the situation in which there are no benthic animals at all, such as commonly occurs after severe physical disturbance, an oil spill, or an episode of anoxia.

5.4 The case of the missing benthos

In the absence of bioturbating benthic animals, and for a given region of the seafloor or depositional basin, variability in the spatial distribution of pollutants will be predominantly in the vertical direction. In this simplest of cases, the concentration of a reactive pollutant subject to microbial degradation is given by the following equation (Berner, 1980; Aller, 1982b):

$$\frac{\partial C_V}{\partial t} = -\frac{\partial}{\partial x}(\omega C_V) + R \tag{5.1}$$

where C_V is the mass of pollutant per volume of sediment; t is time; x is depth within the sediment (origin at the sediment–water interface and positive downward); ω is the burial or sedimentation rate in units of length per time; and R is the biodegradation rate of the pollutant in mass per time per volume of sediment.

If we assume that compaction is negligible (ω is constant as a function of depth in the deposit) and the system is at steady state (the rate of change in C_V at any given depth is zero), then:

$$0 = -\omega\left(\frac{\partial C_V}{\partial x}\right) - kC_V \tag{5.2}$$

where PAH degradation rate is expressed as a first-order rate process and k is the first-order rate constant in units of $1/t$. In general, the assumption of steady state means that depositional inputs, transport within the sediment through, for example, bioturbation, re-suspension, burial and physical mixing by currents, are in steady-state balance and do not vary with time. The effect of this balance is to keep the concentration of PAH at any depth constant in time. Thus in the absence of bioturbation we expect a simple exponential decay in particle-bound PAH with depth. Rearranging equation 5.2 slightly we obtain:

$$\frac{\partial C_V}{\partial x} = \frac{-kC_V}{\omega} \tag{5.3}$$

which indicates that the steady-state PAH concentration gradient is directly proportional to the decay constant (k) and inversely proportional to the sedimentation rate (ω). This means that in environments with high sedimentation rates, such as eutrophic coastal areas, PAH concentration will decrease less rapidly with depth. In practice this has the effect of driving the pollutant deeper into the sediment. Also, the greater the biodegradation rate constant, the steeper the decline in PAH concentration with depth and the less the pollutant tends to penetrate into the sediment. This can be shown directly by solving equation 5.3 to get:

$$C_V(x) = C_{V0}\, e^{\left(\frac{-kx}{\omega}\right)} \tag{5.4}$$

Assuming that all PAH eventually degrades at some depth (although it may not), we can see that the exact decay profile also depends on the initial surface concentration, C_{V0}. In considering the initial surface concentration, we must specifically deal with the physical or hydrodynamical aspects of the ecosystem that act to influence it. This is an example of a clear physical–biological link. The advantage of setting explicit parameters for these linkages is obvious, and is one of the central goals of our approach.

The exact surface concentration will probably fall between the two end-member situations that are possible in aquatic systems: flux dominated, where the surface concentration is controlled by the depositional flux of PAH; and buffered surface concentration, where the surface concentration is maintained at a constant value due to local hydrodynamic conditions. We will designate these two end-member situations as the **flux** and **buffered** conditions respectively. These boundary conditions can have radically different effects on the final sedimentary concentration profiles. The former situation would tend to characterize deeper, less physically disturbed environments where there is a relatively constant and steady rain of sedimenting detritus and particulates. The abyssal depths and sediments near oil platforms constructed in deeper waters may fall into this category. The other depositional extreme will be characterized by shallow environments where the upper surface of the sediment is disturbed by wave and storm activity or tidal currents. Under these conditions, lateral inhomogeneities in depositional supply will be integrated, or buffered, to nearly constant values (Aller, 1982b; Rice and Rhoads, 1989 and references therein).

Consider for a moment only the flux-dominated condition. Two possibilities exist that can have radically different effects on the fate of the contaminant (Rice and Rhoads, 1989). The first situation occurs when contaminant supply to the sediment surface is not coupled to the sedimentation of inorganic particulates. In the case of PAH this would simply mean that inorganic and organic particulate matter sedimentation rates are decoupled. Decoupling could occur if sorbed PAH tended to associate with autochthonous, epibenthic primary production in shallow environments or if organic and inorganic matter deposition rates were decoupled due to physical or biological processes in the water column. Under these conditions the flux of PAH to the seafloor and ω may vary independently. The second situation occurs when sedimentation of inorganic material and the contaminant are coupled. As we will see below, the sediment profiles for PAH will be very different depending on which of these two situations prevails under local conditions. These considerations act to focus thought on the importance of an integrated, multidisciplinary approach to ecotoxicological problem solving.

The above end-member states can be regarded as boundary conditions for equation 5.1 and can be quantified in the following manner. First, we will consider the case in which flux of PAH to the upper layer of sediment is uncoupled from the sedimentation rate of inorganic material. In this case, if there is a constant PAH flux (J_0: μg cm^{-2} s^{-1}) to the sediment surface and if all the incoming PAH are incorporated into the deposit, then the boundary conditions constraining equation 5.1 are:

$$(1)\ x = 0,\ J_0 = \omega\, C_{V0}$$

$$(2)\ x = \infty,\ \frac{\partial C_V}{\partial x} = 0$$

Condition (1) simply defines the input of PAH at the sediment surface as proportional to sedimentation rate whereas condition (2) indicates that at some point infinitely deep into the sediment column, the concentration gradient of PAH will equal zero. However, if the surface concentration of PAH is buffered at some constant value C_{V0} by some combination of lateral dispersion and re-suspension, then we would replace condition (1) above with:

$$(1a)\ x = 0,\ C_V = C_{V0}$$

setting the surface value to a constant determined by the local hydrodynamic regime and pollutant flux conditions. In addition, one can easily convert between volumetric and mass concentration by taking bulk sediment density and porosity into account through the relationship:

$$C_m = \frac{C_V}{\rho\,(1-\phi)} \qquad (5.5)$$

where ρ is the sediment particle density (g cm^{-3}), ϕ is the sediment porosity (volume of water/volume of solid of density ρ), C_V is the volumetric concentration (mass pollutant per cubic centimetre of wet sediment) and C_m is the mass concentration (mass pollutant/mass dry weight sediment).

5.4.1 Constant surface concentration

To illustrate the model, we will first examine the fate of fluoranthene in sediments in the absence of bioturbation by benthic animals. The fluoranthene molecule consists of two fused benzene rings which share a pair of carbon atoms. The fused rings (naphthalene) are bound to a third aromatic ring with which no carbon atoms are shared.

The biological effects of fluoranthene on benthic animals inhabiting sediments are relatively well studied. LC_{50} values measured for the marine benthic amphipods *Rhepoxynius abronius* and *Corophium spinicorne* are in the range of 15–50 μg l^{-1} interstitial water concentration, indicating very high toxicity (Swartz *et al.*, 1990). Overall, the infaunal free-burrowing *Rhepoxynius* was more sensitive to the contaminated sediments. Toxicities tended to decrease with increasing organic matter of the sediment and appeared to be reasonably well predicted when one accounts for partitioning between the dissolved and sorbed phases (Swartz *et al.*, 1990). Concentrations of fluoranthene in contaminated surficial sediments can be in the range of tens to hundreds of parts per million (p.p.m.) (μg g^{-1} dry weight of sediment) (Shiaris, 1989; National Oceanic and Atmospheric Administration, 1989).

Measurements of surficial sediments from the Hudson/Raritan estuary in New Jersey, USA give values for fluoranthene contamination in the region of 4–5 p.p.m. (mass concentration). If we first assume a hydrodynamically buffered surface concentration of 4.7 p.p.m. (National Oceanic and Atmospheric Administration, 1989), we can plot the expected profiles for a range of first-order biodegradation rates by employing equation 5.4 (Figure 5.2). Here we have used a high sedimentation rate often measured in coastal regions (2 cm per year: Aller, 1982b). Note that, as one might intuitively expect, the fluoranthene inventory (I) is strongly dependent on the degradation rate constant. I is simply the depth-integrated area under the profiles in units of p.p.m. centimetres. The rate constants used here may in fact be quite a bit higher than those of natural sediments. If PAH degrades at negligible rates under anaerobic conditions (e.g., Bauer and Capone, 1985), then field profiles will approach the vertical in the absence of fauna and time-dependent variations in input. This clearly highlights the need for degradation rate studies under environmentally realistic conditions. If we were to hold k constant and vary sedimentation rate, the qualitative shape changes in the profiles would remain the same. That is, I would increase with increasing ω, though the effect of changes in sedimentation rate would be somewhat less than that of the degradation rate changes shown in Figure 5.2.

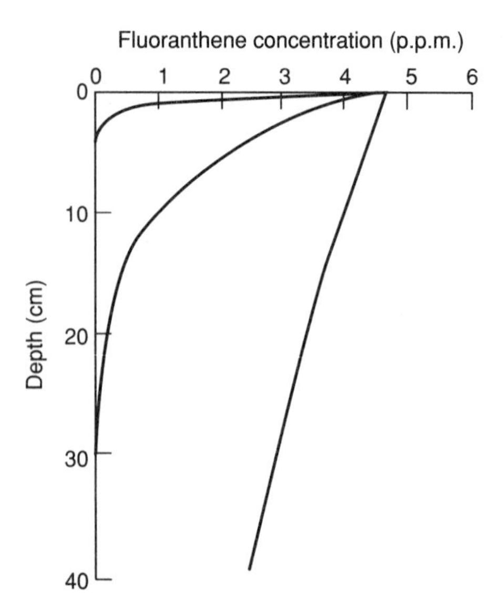

Figure 5.2 Fluoranthene concentration in parts per million ($\mu g\ g^{-1}$ dry weight sediment) as a function of depth for the 'buffered surface concentration' end-member. Profiles are plotted for biodegradation rate constants of 10^{-9}, 10^{-8} and 10^{-7} per second. $C_{m0} = 4.7$ p.p.m. and $\omega = 2$ cm per year ($= 6.3 \times 10^{-8}$ cm s^{-1}).

5.4.2 Flux-dominated condition

Under flux-dominated conditions where pollutant input is uncoupled from clastic sedimentation, the initial surface concentration will be a function of the steady-state flux and follow the relationship: $C_{v0} = J_0/\omega$ (boundary condition (1)). Combining this equation with equation 5.5 we can calculate the initial mass concentration (C_{m0}) as:

$$C_{m0} = \frac{\frac{J_0}{\omega}}{\rho\,(1-\phi)} \tag{5.6}$$

Figure 5.3 depicts fluoranthene concentration as a function of depth under three different steady-state flux and biodegradation rate regimes. The sedimentation rate ω is held constant at 2 cm per year. Not only is the fluoranthene inventory dependent on the flux to the sediment surface, but for a given flux, it is strongly dependent on the biodegradation rate constant as well.

In contrast to the uncoupled situation in which surface concentration of fluoranthene will decrease with increasing burial rate (ω increasing, J_0 constant), if deposition is coupled with inorganic sedimentation, exactly the opposite occurs. Surface concentration increases with increasing ω. In fact, the initial surface concentration (C_{m0}) must be equal to the concentration of

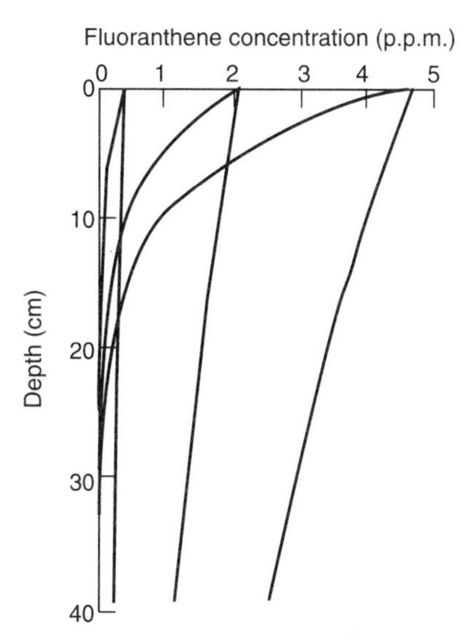

Figure 5.3 Fluoranthene concentration in parts per million (μg g^{-1} dry weight sediment) as a function of depth for the 'steady-state surface flux' condition end-member. Profiles are plotted for three flux values with two k values each ($k = 10^{-9}, 10^{-8}\ s^{-1}$). Model flux values from right to left are: $J_0 = 11.1 \times 10^{-8}$, 5.0×10^{-8} and 1.0×10^{-8} cm s^{-1} and correspond to initial surface concentrations of 4.7, 2.1 and 0.42 p.p.m. respectively.

contaminant on the incoming particulates (C_{mi}: μg PAH g^{-1} dry weight sediment). This intuitive result can be readily shown by noting that $J_0 = C_{mi}\,\rho\omega\,(1-\phi)$, $C_{v0} = J_0/\omega$ and employing equation 5.5.

For the coupled case, the depth profile of fluoranthene will be strongly dependent on both the sedimentation rate and the degradation rate constant. For $C_{mi} = C_{m0} = 4.7$ p.p.m., these dependencies are illustrated in Figure 5.4a and b, and were generated using equations 5.4 and 5.5. Note that, in the absence of bioturbation, and for a given degradation rate, sedimentation rate will have a major impact on the overall concentration profile and depth of penetration of fluoranthene into the sediment. Later, we will show that when the effects of bioturbation by benthic animals are included, ω will be expected to play a role of importance only when sedimentation rates are high.

5.4.3 Estimating interstitial exposure concentrations with the equilibrium partitioning model

Effects of pollutants in sediments are generally determined by their bioavailability rather than strictly by their bulk concentrations. Particle-bound contaminants such as PAH are probably taken up by organisms from solution so that bioaccumulation and toxicities are better correlated with concentra-

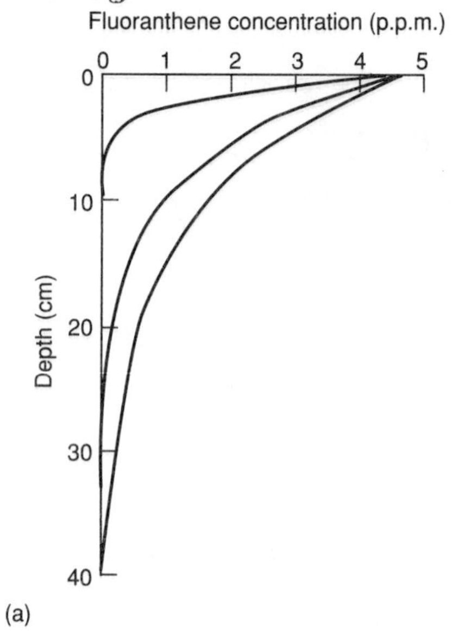

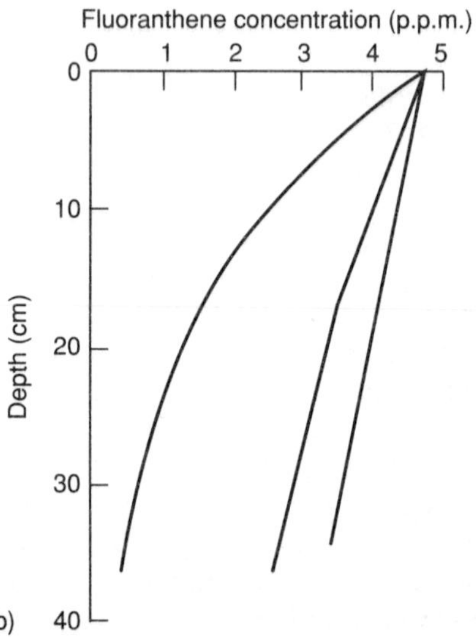

Figure 5.4 Fluoranthene concentration in parts per million (μg g^{-1} dry weight sediment) as a function of depth. (a) The three curves are model plots for sedimentation rates of 1.6, 6.3, and 9.5×10^{-8} cm s^{-1}, where $k = 10^{-8}$ s^{-1}. (b) Curves are model plots for sedimentation rates of 1.6, 6.3 and 9.5×10^{-8} cm s^{-1}, where $k = 10^{-9}$ s^{-1}. Both plots are for the situation in which PAH flux to the sediment–water interface is coupled to ω and $C_{m0} = 4.7$ p.p.m.

tions in interstitial water (Swartz *et al.*, 1990; but see Weston, 1990). Recently, theoretical equilibrium partitioning models have shown promise in improving predictions of the effects of sorbed-contaminants on benthic animals (DiToro

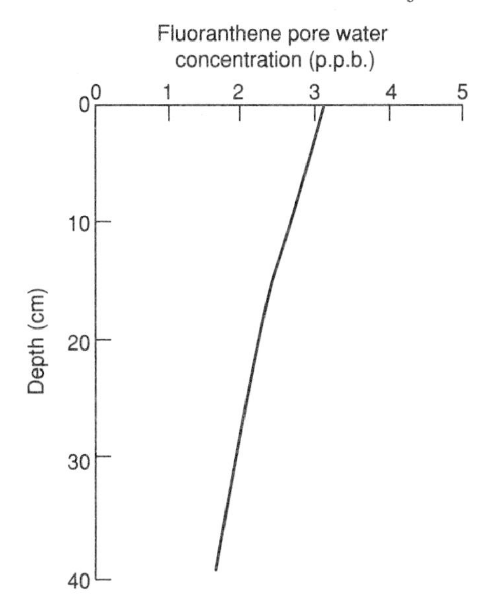

Figure 5.5 Fluoranthene pore water concentration in parts per billion (10^{-9}) as a function of depth in the sediment (μg l^{-1}) calculated using a simple equilibrium partitioning model.

et al., 1989; Swartz *et al.*, 1990). Predictions of interstitial water concentrations of pollutants can be made based on coefficients of partitioning between the particle-bound and dissolved phases. In the case of fluoranthene, which is a non-ionic, hydrophobic, polyaromatic compound; the partitioning between the phases is controlled primarily by the organic content of the sediment (DiToro *et al.*, 1989). The relationship governing fluoranthene partitioning is given by (Swartz *et al.*, 1990):

$$C_{\mathrm{iw}} = \frac{C_{\mathrm{s}}}{K_{\mathrm{oc}} f_{\mathrm{oc}}} \tag{5.7}$$

where C_{iw} is the fluoranthene concentration in the interstitial water (μg l^{-1}), C_{s} is the fluoranthene concentration in the sediment (μg g^{-1} dry weight sediment), K_{oc} is the organic carbon/water partition coefficient for fluoranthene ($\log_{10} K_{\mathrm{oc}} = 4.92$), and f_{oc} is the mass-specific organic carbon content of the sediment (grams of organic carbon per gram dry weight of sediment). Applying this formula to the particle-bound concentration in Figure 5.2 gives a predicted pore water concentration shown in Figure 5.5 (for $k = 10^{-9}$ s^{-1} and an organic content of 1.8%). The figure depicts a smooth drop in PAH concentration paralleling the sorbed concentration from 3.1 p.p.b. (parts per billion taken to be 10^{-9}) at the sediment–water interface to approximately 1.7 p.p.b. at 40 cm. This predicted exposure concentration profile covers the range of approximately 2–3 μg l^{-1}. This range is about 11–20% of the water quality value for chronic effects of 15 μg l^{-1} set by the United States Environmental Protection Agency for fluoranthene (Swartz *et al.*, 1990).

5.5 Effects of bioturbational mixing

The extension of the above arguments to include the effects of biological mixing due to benthic animals is fairly straightforward. The simplest possible effect of animals inhabiting sediments is the homogeneous or nonselective mixing of sedimentary particles during tube construction, feeding and burrowing activity. Animals make it possible for particulate matter to be moved around in a deposit in a manner wholly different from that due to sedimentation and compaction. The problem is that there is no strict physical analogue to this process. In a diverse benthic assemblage, individual particles are capable of complex trajectories as a function of animal activity. For example, particles are subject to local movement and agitation due to burrowing and crawling activities. Animals can also ingest sediment at depth and defecate particles at the sediment–water interface, resulting in advective loops between the surface and depth (Rice, 1986). Or they can feed selectively on both larger (Whitlatch, 1974) and smaller (e.g., Taghon *et al.*, 1978) particles.

One approach to modelling this complexity has been to consider the resulting particle motion as analogous to a mixture of large and small 'particle eddies' (Aller, 1982b). Thus we will consider biological mixing to be analogous to eddy diffusive mixing in hydrodynamics. This can be quantified by defining a new variable, called D_B, as a random mixing coefficient for particulates. For most coastal and deep sea sediments this coefficient is determined largely by bioturbation activity, although purely physical processes can also play a role (e.g., Aller, 1982a). If we again assume a steady-state balance of mixing, burial and degradation within a laterally homogeneous deposit; the change in concentration of fluoranthene with time within the sediment in the presence of mixing can be described by the relation (Cochran and Aller, 1979; Rice and Rhoads, 1989):

$$\frac{\partial C_V}{\partial t} = \frac{\partial}{\partial t}\left[D_B\left(\frac{\partial C_V}{\partial x}\right)\right] - \omega\frac{\partial C_V}{\partial x} - kC_V \tag{5.8}$$

where D_B is in units of $cm^2\ s^{-1}$ (Berner, 1980). All other parameters are identical to the previous examples. Here for simplicity we again assume no compaction and that D_B is independent of depth in the sediment.

5.5.1 *Nature and measurement of* D_B

Empirical values for D_B are typically obtained using naturally occurring radioactive tracers with properties such that they are capable of tracking sedimentary particulates (Cochran and Aller, 1979). One example of a tracer that has been used to measure near-shore processes is thorium-234, which has a half-life of 24.1 days. ^{234}Th is produced by decay of its parent isotope uranium-238. Bhat *et al.* (1969) observed that the ^{234}Th: ^{238}U activity ratio in the near-shore water column decreased to values much less than 1.0 as one approached the coast, that is, ^{234}Th was present at levels below those which would be expected due to radioactive decay of its parent, ^{238}U. Bhat and colleagues interpreted this as indicating that ^{234}Th produced in coastal waters is rapidly extracted from the water column by suspended particulates and transported to the sediment. Later research has supported this interpretation. This was an important result because it means that one can use the excess amount of ^{234}Th over that supported by the decay of ^{238}U as a tracer for following the biological and physical reworking of particles (Aller and Cochran, 1976). Viewed in this way, D_B can be quantified simply as the initial slope of the relationship between excess ^{234}Th and depth in the sediment column (see Cochran and Aller, 1979 for details). The general approach of using naturally occurring radioisotopes to measure physical and biological processes driving vertical particle transport in sediments has been expanded successfully to include a wide range of isotopes of differing half-lives. This approach has allowed measurements to be made over most time scales of interest to benthic ecologists and geochemists (e.g., Aller, 1982b).

Armed with an ability to quantify the mixing parameter D_B, we will again consider both the flux-dominated and buffered surface concentration situations so that boundary conditions (1a) and (2) still apply. With the added complication of biological mixing, we replace condition (1) with:

$$(1b)\ x = 0, -D_B\left[\frac{\partial C_{V0}}{\partial x}\right] + \omega\, C_{V0} = J_0$$

which sets depositional flux equal to mixing and burial into the deposit. For a constant (or depth averaged) D_B, ω and k, the solutions for both flux-dominated (equation 5.8, conditions (1b) and (2)) and buffered surface concentration (equation 5.8, conditions (1a) and (2)) have the same form:

$$C_V(x) = C_{V0}\, e^{\lambda k} \tag{5.9}$$

where

$$\lambda = \frac{\omega}{2D_B} - \sqrt{\left(\frac{\omega^2}{4D_B^2} + \frac{k}{D_B}\right)} \tag{5.10}$$

The initial surface concentrations ($C_V(0)$) for the two conditions are: (1) constant flux to the surface: $C_V(0) = C_{V0} = J_0/\omega{-}\lambda D_B)$ and (2) constant or buffered surface concentration: $C_V(0) = C_{V0}$.

In this case, for simplicity of discussion, we have not specified a functional relationship or depth limit for D_B, although in some cases it is certainly more realistic to do so (see Guinasso and Schink, 1975; Schink and Guinasso, 1978; Matisoff, 1982; Aller, 1982b for non-pollutant examples).

5.6 Linking fate and effect through the biological mixing parameter D_B

5.6.1. PAH depositional flux coupled to sedimentation rate (ω)

How does the biological mixing of sediment affect pollutant degradation and transport when flux to the sediment surface is coupled to total sedimentation rate (ω)? And, more to the point, how can we use this information to begin to link fate and effect? To investigate these questions we will employ equations 5.9 and 5.10 using the appropriate boundary conditions. We first note that ω will have a relatively insignificant effect on the pollutant depth profile with respect to D_B if (Rice and Rhoads, 1989):

$$\omega << \sqrt{4kD_B}.$$

Because we will be investigating coastal or estuarine areas with relatively high sedimentation rates; we will include ω and examine its importance relative to D_B. Coastal environments such as Long Island Sound or the lower Hudson River, USA have sedimentation rates in the order of 1–2cm per year. As we demonstrate below, sedimentation rates this great can noticeably influence pollutant depth profiles.

Figure 5.6 illustrates the effect of infaunal bioturbation on the depth distribution of fluoranthene. Model values are given in the caption. In this example we have used a fairly rapid degradation rate constant of $10^{-7}s^{-1}$; but the qualitative conclusions remain the same for slower degradation rates. Here we have used boundary values for D_B such that the range of previously measured values spans coastal to deep-sea environments (Table III from Aller, 1982b; but see also Table I in Matisoff, 1982). As with increasing sedimentation rate, the overall effect of increased animal activity is to drive the pollutant deeper into the sediment. This follows naturally because the sorbed pollutant is mixed more deeply into the sediment before it can decay. However, note that when pollutant flux is coupled to ω, the surface or initial concentration will be constant if ω is constant. This situation is functionally identical to the previous situation of the 'hydrodynamically dominated' environments in which the surface concentration is buffered at some constant value (cf. Figure 5.3 and 5.7). This has extremely important consequences for the sedimentary inhabitants and perfectly illustrates the type of situation in which pollutant fate and effect can become strongly linked at the community or ecosystem level. For example, under these conditions, reworking activities of the animals act to increase the total amount of pollutant stored or incorporated into the sediment. This will increase the pollutant exposure to the animals themselves, creating a feedback loop between pollutant fate and effect. *This feedback loop will be quantified by the functional relationship between pollutant exposure and bioturbation rate.*

The total amount of pollutant incorporated into the sediment as a function of reworking activity can be conveniently represented as the benthic inventory,

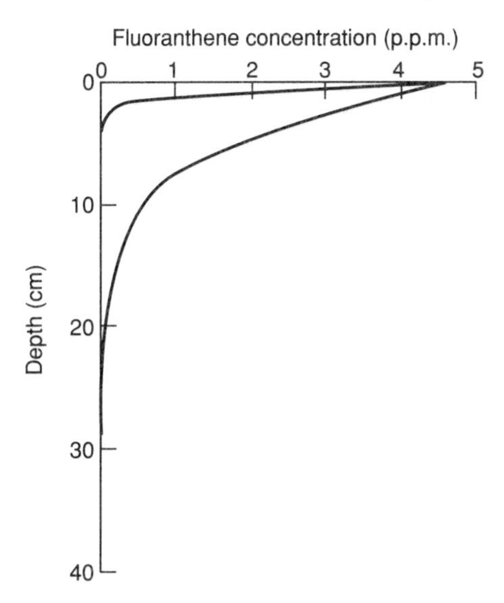

Figure 5.6 Fluoranthene concentration in parts per million (μg g^{-1} dry weight sediment) as a function of depth. Flux of fluoranthene is coupled to bulk sedimentation rate. Curves are for values of D_B of 6×10^{-9} cm^2 s^{-1} ($I = 1.4$ p.p.m. cm) and 600×10^{-9} cm^2 s^{-1} ($I = 11.8$ p.p.m. cm), where $k = 10^{-7}$ s^{-1} and $\omega = 10^{-8}$ cm s^{-1}.

The total amount of pollutant incorporated into the sediment as a function of reworking activity can be conveniently represented as the benthic inventory, I, which was defined above. We can quantify I for the curves in Figure 5.6 for the top 20 cm of sediment by numerically integrating the expression:

$$\int_0^{20} C_{m_i} e^{\lambda x} \, dx \tag{5.11}$$

for both curves in Figure 5.6. All parameters are as defined previously. Thus for the more slowly reworked sediment (i.e. $D_B = 6 \times 10^{-9}$ cm^2 s^{-1}) $I = 1.4$ p.p.m. cm, whereas for the more rapidly bioturbated sediment ($D_B = 600 \times 10^{-9}$ cm^2 s^{-1}) $I = 11.8$ p.p.m. cm. *Under the above conditions rapidly bioturbating infauna increase their own exposure to fluoranthene (equation 5.7) by a factor of more than eight.* Note that one views pollutant fate–effect linkages as the feedback loop between contaminant inventory and D_B and that pore water and adsorbed inventories can be readily interconverted using equation 5.7. Population density and/or biomass are known to influence D_B, but the details of the causal relationship are unexplored (Matisoff, 1982). To the extent that there is linkage, there will be a feedback relationship between exposure concentration and population density or biomass. Clearly, increased understanding of PAH fate and effect in sediments requires quantification of the relationship between biological reworking (and its subcomponents) and pollutant exposure.

We can illustrate this with the following conceptual approach. If we assume that effect will be closely related to interstitial exposure concentration, then based on the above discussion we note that the ultimate effect of a particle-associated, biodegradable pollutant such as fluoranthene will be a function of physicochemical *and* biological variables. These relationships can be conceptualized by noting that exposure or effect will be some function of benthic inventory and reworking activity. D_B is a function of both the biological system under consideration and the effect of the pollutant on that system. Pollutant inventory will be a function of both physical and biological variables. An additional complication arises from the fact that because inventory is partially a function of the biology of the system, it will also be influenced by pollutant effect. In order to frame these thoughts within a more concise format, we propose the following conceptual fate–effect coupling for particulate or particle-associated biodegradable pollutants:

$$Exposure = f\{D_B\ (Effect,\ biology),\ I\ (Effect,\ biology,\ physics,\ chemistry)\}$$

where we operationally define *effect* as any pollutant-caused perturbation that acts to change the value of D_B. The parameters *biology, chemistry* and *physics* simply represent the sum total of the relevant variables (yet to be determined) within each respective realm. Focusing more closely on the relationship defining D_B, the biological component can in turn be reduced to the individual and population-level parameters of which it is composed. In this case $D_B = f\{$*Effect, individual level biology (body size, growth rate, fecundity, ...), population level biology (biomass, density, ...)*$\}$. This conceptualization has several advantages. The first is the explicit characterization of the hierarchical nature of the relationships among the key parameters. The second and related advantage involves the codification of the fundamentally recursive nature of the feedback relationship between pollutant fate and effect in marine sediments. Viewed in this way, these linkages can provide the basis for the next generation of experiments, models and improved predictions. The value of this approach lies in the fact that the parameters considered important to PAH fate and effect are explicitly stated. Once the potentially important variables are clearly defined, experiments can be designed to evaluate their relative importance within the framework sketched out above.

As a partial step in this direction, we can model the relationship between fluoranthene inventory and D_B. That is, given the model relationships and assumptions we have so far defined, how will inventory be related to D_B? We can plot this relationship by integrating equation 5.11 for a range of D_B values. Figure 5.7 depicts the relationship between fluoranthene inventory and bioturbation rate over the range of rates considered above (for constant $\omega = 10^{-8}$ cm s^{-1} and $k = 10^{-7}$ s^{-1}). Inventory increases curvilinearly by a factor of more than 10 with increasing D_B. Notice that the greatest absolute change in I per unit change in D_B is predicted to occur at the lowest reworking rates. This feature has important implications with regard to higher level biological

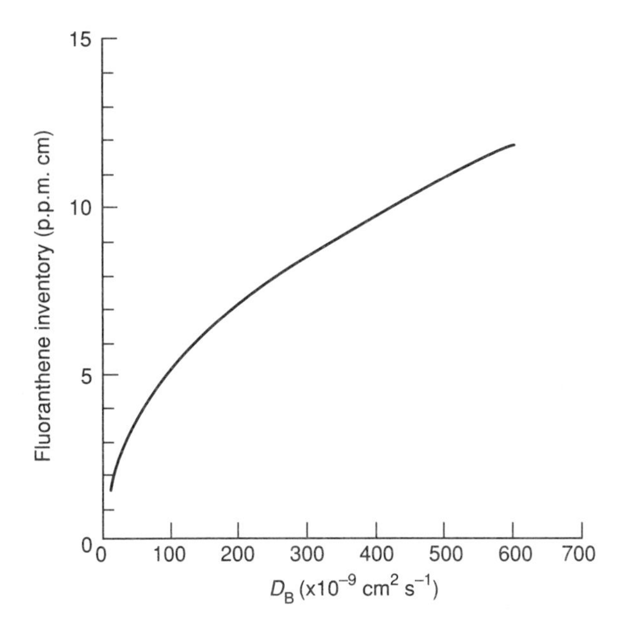

Figure 5.7 Fluoranthene inventory versus the biodiffusion coefficient for particulates (D_B). Flux of fluoranthene is coupled to bulk sedimentation rate.

processes. For example, pollutant inventory (and therefore the feedback loop between fate and effect) should be most sensitive to changes in bioturbation rates at the left-hand side of Figure 5.7, where the slope is the greatest. As a general rule, low values of D_B would be expected under deep-sea conditions where animal activity is kept low by food limitation. We might also expect low bioturbation rates in the coastal zone during the early stages of ecological succession. If the local fauna have been wiped out due to eutrophication or some other physical or pollutant-induced disturbance, the early recolonization will be associated with shallower benthic reworking and possibly lower values of D_B (e.g., Rhoads, 1974). To the extent that the above relationship holds, deep-sea or early successional communities would be expected to be influenced to a relatively greater degree by fate–effect couplings involving changes in D_B. This is due in part to the relationship of D_B to population-level variables such as biomass or population density (Matisoff, 1982). We discuss this relationship as a possible avenue for linking population-level parameters to pollutant fate and effect below.

To provide a more intuitive feel for the relative importance of bioturbation and sedimentation rates for this example, we have plotted fluoranthene inventory as a function of both of these parameters in Figure 5.8. The three plotted sedimentation rates are typical of coastal areas and range from approximately

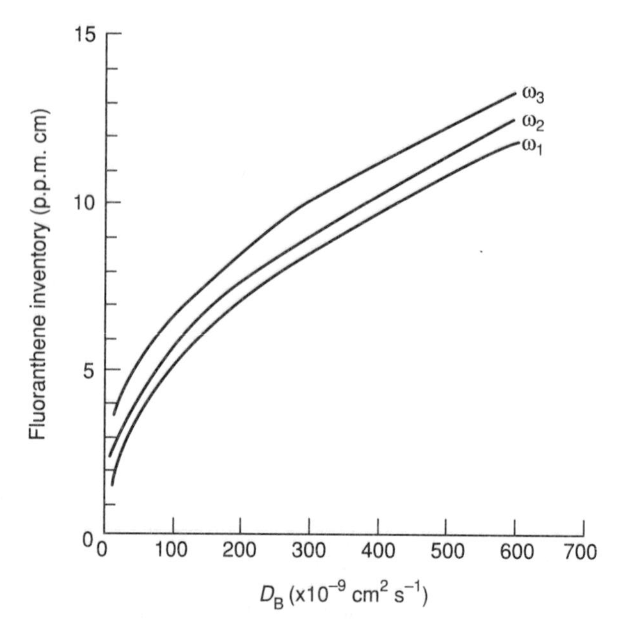

Figure 5.8 Fluoranthene inventory versus the biodiffusion coefficient for particulates (D_B) for sedimentation rates (ω) of $\omega_1 = 1.0, 10^{-8}$, $\omega_2 = 3.0 \times 10^{-8}$, $\omega_3 = 6.3 \times 10^{-8}$ cm s^{-1}. Flux of fluoranthene is coupled to bulk sedimentation rate.

0.3–2.0 cm per year. Note that even in areas of fairly high deposition, biological reworking will be the dominant factor controlling pollutant inventory.

5.6.2 Situation in which PAH flux is uncoupled from bulk sedimentation rate

Uncoupling fluoranthene flux from bulk sedimentation rate has very important consequences with regard to both the surface concentration and depth profile. For a given biodegradation (k) and bulk sedimentation rate (ω), changes in benthic reworking rates will have two major effects. First, surface concentrations will decrease with increasing reworking rates (Figure 5.9; model parameters given in the figure caption). In fact, surface concentration decreases exponentially with increasing D_B. This follows from the general principle that increased reworking rates tend to cause newly deposited material to be transferred to depth more rapidly. The second important result is that if fluoranthene flux to the surface (ω), and k remain constant, then changes in reworking rate alone will have no effect on the total benthic inventory (Figure 5.9). For both curves depicted in Figure 5.9, the fluoranthene inventories are 3.0 p.p.m. cm. The conclusion that follows is that the local physical and biological conditions determining the surface boundary concentration will be extremely important ecosystem-level parameters determining the fate and effects of particle-associated pollutants such as PAH.

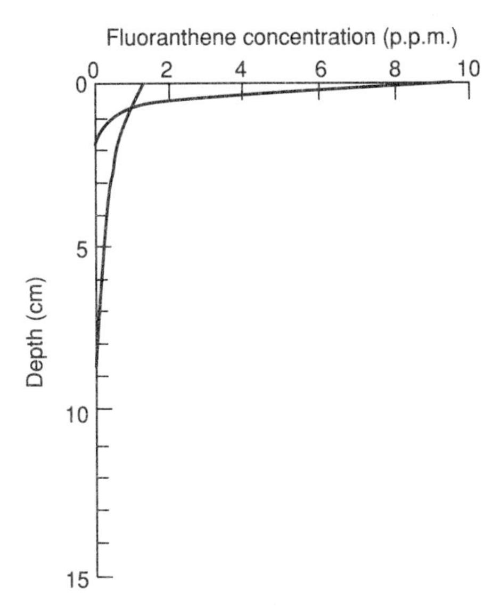

Figure 5.9 Fluoranthene concentration in parts per million (µg g^{-1} dry weight sediment) as a function of depth. Flux of fluoranthene is decoupled from bulk sedimentation rate. Model curves are shown for $D_B = 6 \times 10^{-9}$ and 600×10^{9} cm^2 s^{-1}; where $J_0 = 1.11$ µg cm^{-2} s^{-1}, $\omega = 10^{-8}$ cm s^{-1} and $k = 10^{-7}$ s^{-1}. The inventory down to 20 cm depth is 3.0 p.p.m. cm^2 for both curves.

Changes in reworking rates can also have important consequences with regard to pollutant exposure that depend on the depth in the sediment at which the animals feed or live. Figure 5.10 is a plot of fluoranthene concentration as a function of reworking rate at three sedimentary horizons: 1.0, 1.5 and 2.0 cm. Sedimentation rate, PAH flux and biodegradation rate are the same as in Figure 5.9. Two features of this graph are significant. The first point to note is that there are fluoranthene concentration maxima. The peak can be more or less broad but nevertheless there are values of D_B which will maximize exposure for any given set of conditions. Also, there will be a particular depth for which the peak is greatest. For any environment, the depths for which exposure is high or the peaks are narrow will depend intimately on the prevailing biological and physical conditions. Specifically, the relationship depends on J_0, ω, k and whether or not PAH flux is coupled to bulk sedimentation rate. Note also that under any given set of conditions, there will be a region in which changes in reworking rates will have a relatively great effect. For example, reworking rates in the range zero to approximately 200 cm^2 s^{-1} will have the greatest effect on exposure concentration in this particular example. This suggests that reworking rate changes during recolonization or early succession (such as occur after pollutant input or natural disturbance) can potentially have a great effect on subsequent exposure.

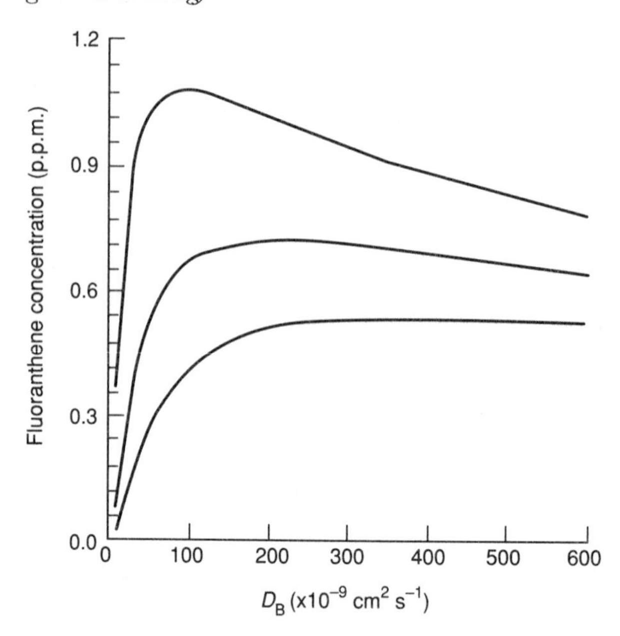

Figure 5.10 Fluoranthene concentration in parts per million (μg g^{-1} dry weight sediment) as a function of the biodiffusion coefficient for particulates (D_B) for three near-surface depth horizons: $z = 1.0$ cm (top curve); $z = 1.5$ cm (middle) and $z = 2.0$ cm (bottom). All model curves are for $\omega = 10^{-8}$ cm s^{-1}, $k = 10^{-7}$ s^{-1} and $J_0 = 1.11 \times 10^{-7}$ μg cm^{-2} s^{-1}. Flux of fluoranthene to the sediment surface is decoupled from bulk sedimentation rate (ω).

5.7 Coupling fate and effect to the population level of organization

The most straightforward approach to linking fluoranthene fate and effect to higher levels of biological organization involves the biological/physical mixing coefficient, D_B. Measured or estimated values of D_B for marine sediments have been found to range between 10^{-4} and 10^{-10} cm^2 s^{-1} (Schink *et al.*, 1975; Matisoff, 1982). Matisoff suggested that D_B can be thought of as an empirically measured 'net biodiffusivity', which is the product of what he termed a 'biomass density biodiffusivity', D_β (cm^4 g^{-1} s^{-1}), and the biomass density of organisms, M (g dry weight cm^{-2}):

$$D_B = D_\beta M. \qquad (5.12)$$

In this case one can simply normalize D_B by biomass density to get D_β, the biomass density diffusivity. Thus D_β is a proportionality constant which relates benthic biomass to the mixing coefficient, D_B. Matisoff (1982) noted that D_β was approximately 90% less variable than D_B. If biomass accounts for the great majority of the variation in the reworking parameter, then so also must the individual- and population-level parameters such as organism size and population density that combine to influence biomass. These parameters in turn will be influenced by critical factors such as individual growth and reproductive rates. Of the numerous investigations that have measured D_B, only a very few have also measured infaunal biomass (Matisoff, 1982, his Table II). We can use this limited data set ($n = 10$ data pairs) to obtain an empirical relationship between D_B and M and thus test the validity of equation 5.12. Note that if equation 5.12 as proposed by Matisoff actually holds, we would expect a power curve with an exponent not significantly different from 1.0. There is a good deal of scatter, but the data nevertheless follow a power curve fairly well (Figure 5.11). The least-squares fit is significant ($R^2 = 0.58$, $n = 10$, $P = 0.010$) and the power function is given by:

$$D_B = D_\beta M^{0.52 \pm 0.153} \tag{5.13}$$

where $D_\beta = 5.77 \times 10^{-6}$ ($cm^4\ g^{-1}\ s^{-1}$) and the error for the exponent is plus or minus one standard error. The power curve exponent is dimensionless. Given the magnitude of the standard error, the exponent is indeed less than 1.0 (t-test, $P < 0.05$), invalidating the simpler relationship given in equation 5.12. Thus the biological mixing coefficient appears to be roughly proportional to the square root of the biomass density:

$$D_B \propto \sqrt{M} \tag{5.14}$$

where D_β is the constant of proportionality. This empirical result has important implications. It suggests that, as a general rule, biological mixing per unit biomass decreases rapidly with increasing biomass. This tentative result is based on extremely scant data and it is very important to know to what degree the above relationship holds. This relationship is fundamental with regard to attempts to couple fate and effect at the population level. This is one sense in which D_B can be considered a population parameter. For example, relating D_B to successional stage, or any community classification scheme based on species composition would greatly aid in improving predictions of exposure and fate of pollutants under natural field conditions.

As our knowledge in this area increases, we will no doubt find many species-specific differences from any general relationship that will be related to specific life history characteristics or adaptations. For example, the highest weight-specific deposit feeding rates yet measured are for the infaunal polychaete, *Scoloplos* spp. This worm is capable of ingesting up to 120 times its own dry weight in sediment per day (D. Rice, unpublished; cited in Lopez and Levinton, 1987). In addition, this reworking of particulate matter is strongly directional in nature, not random as the definition of D_B suggests. *Scoloplos* spp. ingest sediment at depth and egest it at the sediment–water interface; thus creating an advective loop between the sediment surface and the feeding depth. Rice (1986) has demonstrated that this feeding behaviour is capable of subducting reactive organic matter much more rapidly than would otherwise be the case, suggesting that high weight-specific feeding rates may be an adaptation allowing *Scoloplos* to compete more effectively with sediment surface-dwelling

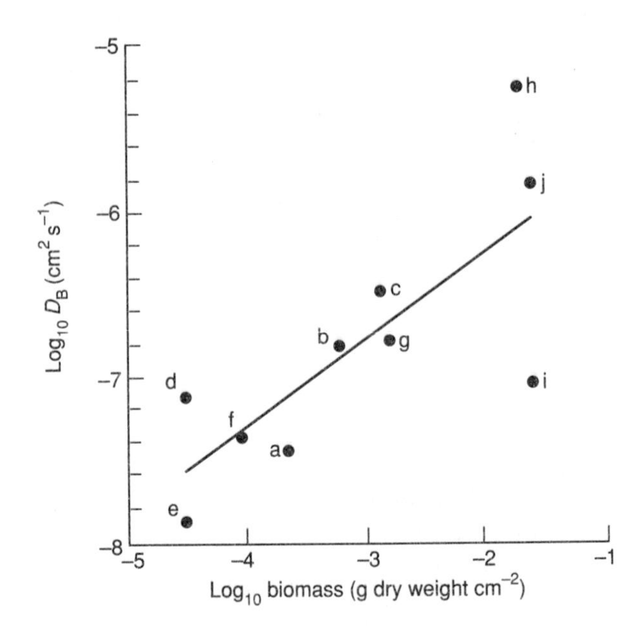

Figure 5.11 Log–log plot of the biodiffusion coefficient for particulates (D_B) versus biomass density. See text for equation. Data from Matisoff (1982). The organisms and locations of data collection are: (a) *Yoldia limatula*–Buzzards Bay, USA; (b) *Y. limatula*–Long Island Sound, USA; (c) *Y. limatula*–Long Island Sound, USA; (d) *Pectinaria gouldii*–Barnstable harbour, USA; (e) *P. gouldii*–Barnstable harbour, USA; (f) *Limnodrilus* and *Tubifex tubifex*–Messalonskee Lake, USA; (g) *Pontoporeia hoyi*–laboratory; (h) *T. tubifex*–laboratory; (i) *T. tubifex*–laboratory; (j) *T. tubifex*–laboratory.

fauna for more digestible food. Communities dominated by *Scoloplos* or other so-called conveyor belt species might be expected to build up relatively large inventories of PAH or other particle-bound pollutants. This is in addition to any considerations regarding buffered or flux-dominated interfacial conditions. Communities dominated by conveyor belt species are generally regarded by marine ecologists as relatively late-successional or low-disturbance communities. This is because these larger, deep-feeding, slow-growing animals cannot maintain viable populations under conditions where the benthos is wiped out on time scales shorter than their relatively long life histories. In contrast, the biomass of early successional communities is often dominated by small, fast-growing species that tend to live and feed near the sediment–water interface, generally in the upper 1–2 cm (e.g., Rhoads, 1974; Pearson and Rosenberg, 1978). The reworking due to these early succession communities is often less directional in orientation as well. The above considerations suggest that later successional communities may often increase their own exposure to particle-bound pollutants and may also tend to have a greater impact on pollutant fate (inventory, profile, degradation rate) than their early successional counterparts.

5.7.1 Geochemical hotspots and bioturbation

From the above discussion it is clear that natural communities exhibiting relatively high bioturbation rates will function as local 'geochemical hotspots' with regard to particle-sorbed pollutant diagenesis (e.g., after Aller, 1982). Especially under conditions of buffered surface concentrations, pollutant inventory (and thus exposure) will increase strongly as a function of reworking rate (Figure 5.7). This means that those communities or assemblages with high reworking rates will tend to scavenge or accumulate pollutants. The natural outcome of this process will be a spatial and temporal mosaic of pollutant exposure and loading which is strongly correlated with local reworking rates. The rates in turn will be a function of community structure and dynamics. The final link in the pollutant fate and effect dynamic comes about as these reworking rates are influenced, to some as yet unknown degree, by the exposure levels they have created. Thus, in addition to information on the local hydrodynamic regime, meaningful studies of pollutant fate and effects in marine sedimentary systems must also include biological data on the local community. These data should probably include species abundances, depth of bioturbation and other key characteristics that may be relevant to determining a functional relationship for D_B. The critical challenge then, is to determine the functional or feedback relationships among the most important individual and population-level parameters determining D_B. Thus we propose a central role for biological reworking in coupling particle-bound pollutant fate and effect which is illustrated conceptually in Figure 5.12.

An important area of research in the future will involve the effects of interactions among pollutants. In terms of the effects of additional pollutants on particle-sorbed contaminants, we can again view pollutant interactions

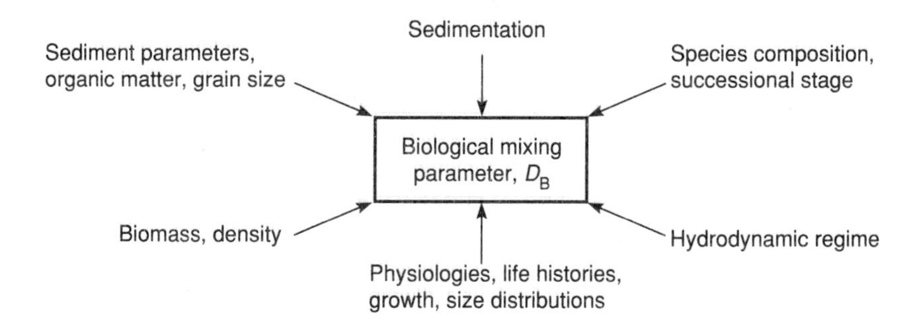

Figure 5.12 The central role of biological reworking in coupling the fate and effects of biodegradable pollutants in marine sediments.

through their effect on the mixing or reworking parameter. For example, increased organic loading to the marine environment due to poor sewage treatment or runoff from agricultural fertilizers will have several important consequences. These include:

1. increased particulate organic matter; (and therefore PAH) deposition at the sediment water interface;
2. community species composition changes due to the different organic matter and/or PAH flux;
3. hypoxia or anoxia due to aerobic respiration of the additional organic matter;
4. the resulting influence of the above factors on D_B.

This emphasizes the fact that quantifying the relative importance and feedback relationships among these factors is crucial to making ecosystem-level predictions concerning interactions between eutrophication and PAH fate and effects.

5.8 Quantifying the effect of bioturbation on biodegradation

5.8.1 Depth-integrated biodegradation potential

We can further extend the above approach and increase its ecotoxicological usefulness by examining the relationship between reworking and pollutant biodegradation. The biodegradation potential of a given volume of sediment depends on both the amount of pollutant present (C_V) and that pollutant's inherent biodegradability under the prevailing local conditions (k). As we will illustrate quantitatively below, it will also depend on bioturbation by the local infauna. Adapting an approach analogous to that of Rice and Rhoads (1989) for reactive organic matter, we can define the pollutant biodegradation potential (BP) for an arbitrarily small volume of sediment, as a volumetric reactivity:

$$BP = -\frac{\partial C_V}{\partial t} = kC_V \tag{5.15}$$

With this definition, the biodegradation potential can be thought of as simply the volumetric rate of biodegradation (units; μg PAH cm^{-3} sediment s^{-1}). At steady state, the depth-integrated biodegradation potential (IBP) of a layer of sediment (from x_1 to x_2) will be equal to the net flux of biodegradable pollutant into that interval. This net flux is expressed simply as:

$$IBP\ (x_1,x_2) = J_0(x_1) - J_0(x_2). \tag{5.16}$$

Equation 5.16 states that a degradable organic pollutant that enters the layer at depth x_1 but does not exit at depth x_2 has been biodegraded. Note that the units of IBP flux are the same as those for benthic respiration measurements in general (i.e. oxygen consumption), and are mass per area per time.

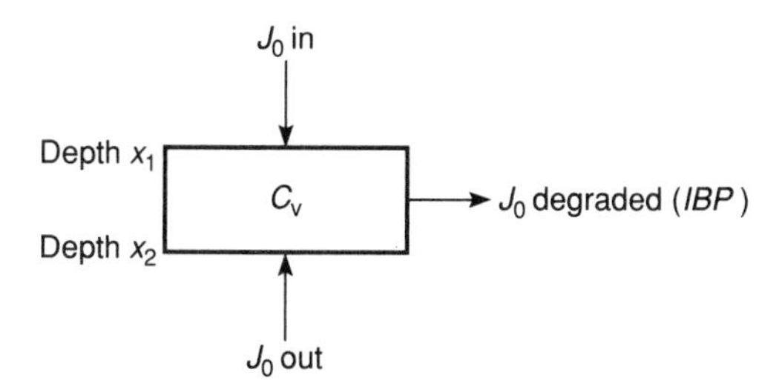

Figure 5.13 The integrated biodegradation potential (*IBP*) defined as the 'biodegradative flux' of an organic toxicant within any arbitrarily thick layer of sediment. (*IBP* units: µg pollutant degraded $cm^{-2}\ s^{-1}$). See text for detailed discussion.

Thus we can conveniently think of *IBP* as the biodegradative flux occurring between any two arbitrary depth intervals. This concept is illustrated diagrammatically in Figure 5.13.

For a given burial rate (ω) and flux of pollutant to the sediment–water interface; *IBP* will be a function of depth, the biodegradation rate constant and the biodiffusion coefficient (D_B) and follow the expression:

$$IBP = [C_v(x_1,D_B) - C_v(x_2,D_B)]\ \sqrt{kD_B} \qquad (5.17)$$

where $C_V(x,D_B)$ is calculated using equations 5.9, 5.10 and the appropriate boundary conditions.

As a final illustration of our approach, we will use equation 5.17 to examine the importance of D_B in the biodegradative flux of fluoranthene in the top 10 cm of the sediment column. Figure 5.14 is a plot of *IBP* for the top 10 cm of sediment for two sedimentation rates ($\omega = 1.0 \times 10^{-8}$ and 3.17×10^{-8} cm s^{-1}). In this example the flux of fluoranthene to the sediment is uncoupled from burial rate (where $J_0 = 1.11 \times 10^{-7}$ µg $cm^{-2}\ s^{-1}$ and $k = 10^{-9}\ s^{-1}$). Both sedimentation and reworking rates can have a large influence on *IBP* in the upper sediment layer. Note that for a given biodiffusion coefficient, increasing burial rate will decrease the overall degradative flux in the surface layer. This is because the faster material passes through the layer, the less time it has to degrade. This effect may be important in some coastal areas that combine high sedimentation rates with shallow reworking depths. Such environments include highly eutrophic areas maintained in early successional stages by organic loading and associated hypoxia (e.g., Pearson and Rosenberg, 1978). If compounds are degraded only in the upper zone of biological reworking, then the total amount of pollutant degraded before ultimate burial could be significantly influenced by sedimentation rate. For fluoranthene degradation in the upper 10 cm, the longer the residence time (as reflected in the lower sedimentation rate), the greater the overall effect of reworking activity (Figure 5.14). The relative importances of D_B and ω will vary with both their own absolute values and the

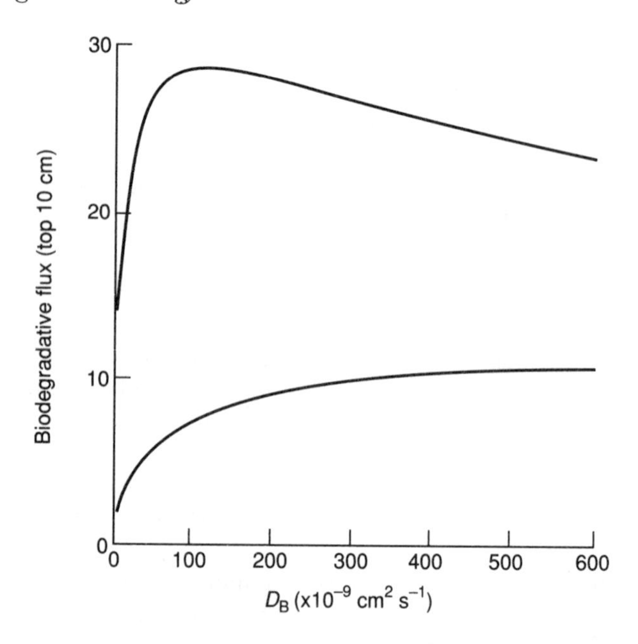

Figure 5.14 Model curves of biodegradative flux (*IBP*) as a function of the biodiffusion coefficient for particulates (D_B) in the top 10 cm of the sediment column. Model curves are for $\omega = 1.0 \times 10^{-8}$ cm s^{-1} (top) and $\omega = 3.2 \times 10^{-8}$ cm s^{-1} (bottom) where $J_0 = 1.11 \times 10^{-7}$ µg cm^2 s^{-1} and $k = 10^{-9}$ s^{-1}. The flux of fluoranthene to the sediment surface is decoupled from bulk sedimentation rate (ω).

depth interval of interest. Thus, in addition to strongly influencing particle-reactive pollutant inventories, biological reworking can also have a very significant effect on total biodegradation in aquatic sedimentary systems.

5.9 The science gap

Unfortunately, the geochemical processes and relationships that we have explored above in great detail, and which are widely employed by oceanographers, seem to have gone completely unnoticed at the administrative and regulatory level. As a case in point, we again cite the OECD guidelines. The text below is taken from the section entitled: *Summary of Considerations in the Report from the OECD Expert Group on Degradation and Accumulation* (Organization for Economic Cooperation and Development, 1981). Within the subsection on *Behaviour in Soils and Sediments* we have: 'The study of the fate of chemicals in soils or sediments is of particular importance since contamination can be long lasting and difficult to reverse due to limited opportunity for mobilisation and dilution.' So far so good, but the next paragraph continues with: 'When performing tests for biodegradation, or the adsorption coefficient, normally, **the structure of the soil is destroyed**, for example by sieving or washing, while texture is maintained. **This is of minor importance for the determination of the degradability potential**' (emphasis added).

The discovery of this passage suggests that lack of incorporation of basic biogeochemical knowledge into regulatory guidelines is a very widespread and serious problem. Clearly, sediment structure, not to mention the associated infauna, are of central importance in determining biodegradation in aquatic depositional environments. Even the texture of marine sediments may be changed by sieving and the concomitant destruction of the physical and geochemical milieu (Johnson, 1974; 1977; Watling, 1991). The lack of incorporation of recent advances in knowledge into the OECD guidelines surprised us. This is dangerous not only because current environmental regulations are based on it, but future research is guided by it as well (Chapter 6).

As we have illustrated with the example above, there is at present a large gap between the acquisition of scientific knowledge and its incorporation into regulatory guidelines. The lack of assimilation of recent (and not so recent) developments in the ecology and biogeochemistry of marine sediments provides only one example. Many others can be found. Presently existing knowledge is not being used to the fullest extent possible. The present OECD guidelines for the testing of chemicals in sediments reflect the scientific understanding of 20 years ago. The reasons for the gap are no doubt complex, but two of the potentially most important factors are sketched below.

There seems to be a general narrowness in the training of contemporary ecotoxicologists and environmental regulators. This inhibits the ability to recognize and follow important developments in the study of more complex problems. The most important of which are probably in the areas of ecology, biogeochemistry, evolutionary biology, statistics and mathematics. Because of a lack of appropriate tools, many workers may suffer from an inability to address (or even appreciate) the important problems we currently face. Clearly the solution to this problem is improved and broadened graduate training (see Chapter 6). Unfortunately, there appears to be a general trend towards increasing specialization as reflected in the development of specialized graduate programmes in 'ecotoxicology'. We view this trend as unproductive and potentially damaging. Because there will be a substantial time lag between improvements in training and the influx of better qualified scientists and regulators into the ecotoxicological community, it is imperative that these improvements be made as rapidly as possible.

Secondly, because of its obvious societal importance, there is often an unfortunate propensity for many of the research questions in ecotoxicology to be directed by nonscientific forces. The strong influence of political and administrative bodies often constrains much of the research to be less than state of the art. The current situation in Denmark is a perfect case in point and the Danish approach to the funding of environmental research has been criticized recently by Depledge (1992). In Denmark, a great deal of ecotoxicological research is funded through the Danish Environmental Protection Agency (DEPA), which currently functions solely as an administrative and regulatory body. However, DEPA holds the purse strings to a large fraction of the potential funding available to Danish ecotoxicologists. The personnel evaluating the grant proposals are not active scientists and thus are not in the best position to judge the scientific quality of the proposed research. However, when a grant proposal is submitted to DEPA, evaluations and funding decisions are often made in the complete absence of valid peer review. From this directly follows the unfortunate situation in which the research is very often of mediocre quality and fails to improve our understanding of toxicants in natural systems. Similar problems in the American approach to funding can be seen in the recent trend by Congress to favour projects benefitting their home districts at the expense of peer reviewed proposals. It also appears that Congress is taking an increasing role in specifying the goals of federally funded research. As Abelson (1992) noted, congressmen have obvious limitations as directors of research, not the least of which is that their time horizon is too short to guide the nation in achieving long-term scientific goals. Despite its imperfections, a scientifically valid peer review process offers the best hope for ensuring that limited funds are directed towards high quality, rigorously defined research projects.

5.10 Summary and conclusions

Modern ecotoxicological research demands an integrated approach to problem solving at levels of complexity that have been neglected in the past. The present unfortunate dependence of regulatory criteria on simplistic laboratory assays of toxicants is a direct result of our ignorance of the dynamical aspects of toxicant effects in natural systems. A great deal of effort and expense has been focused on biological effect and pollutant fate studies in recent years, yet we are far from a predictive science of chemical toxicity at the level of the ecosystem or natural community. Our ignorance stems directly from the fact that past studies have tended to deal exclusively with the fate **or** effects of pollutants **and** have been directed at levels below the population or ecosystem. This failure to approach ecotoxicological problems in an integrated fashion is partly due to historical constraints and partly to the inherent difficulty of the problems themselves.

As an illustration of the general type of approach we feel would be useful in the future, we took a deliberately oversimplified but integrated look at the fate and effects of the polyaromatic hydrocarbon, fluoranthene, in recently deposited marine sediments. Stepping through this example in detail allowed us to isolate, *a priori*, what should be the most critical biological, chemical and physical parameters influencing the ultimate fate and biological effects of fluoranthene in the marine environment. Though the concern at the heart of all ecotoxicological studies must ultimately be the biological effects of toxicants, the illustrative example graphically demonstrates that purely physical and chemical considerations are often crucial to a full understanding of the toxic effects of chemicals in nature. Future investigations and models must be developed at levels of complexity much greater than has been the case for past research. These investigations will involve integrating the critical physical and biological components of toxicant action at the level of the population and ecosystem. There is little to be gained by additional rounds of overly simplistic fate or effect studies carried out in isolation.

Finally, we suggest that there is often a large gap between currently available scientific knowledge and environmental legislation. This leads to the development of antiquated regulatory guidelines and the initiation of low-quality lines of research. This unfortunate situation stems from a lack of widespread, broad-based training and the dominance of political and administrative considerations which often act to drive research.

6 *Ecotoxicology: Past, Present and Future*

In a few centuries cultural evolution has produced the most persistent and pernicious notion ever to afflict humanity. It is the belief that what we do today is always better than what we did yesterday, that all forms of 'progress' are desirable, inevitable, and irreversible.
R. Ornstein and P. Ehrlich, 1989

Civilization is the creation of optimistic minds.
Robert Ricklefs, 1979

There have been two primary goals in the development of ecotoxicology since its inception in the late 1960s. The first was to increase the understanding of pollutant fate and effects in the environment, and the second was to develop ecotoxicological test methods in order that pollutant impacts can be predicted, detected and controlled. Failure to recognize the disparity between these goals has been at the heart of much of the controversy regarding the appropriate scope and direction of research efforts in ecotoxicology. In the short-term, ecotoxicology must address the demands posed by the release of chemicals into the environment **today**. This means that we need a suite of effective test methods that will quantify the potential risks associated with chemical use as accurately as available knowledge permits. However, to focus all, or even the bulk of our efforts on the development of test systems would be a grievous mistake. In the long term, our only hope of effectively dealing with the vast array of chemicals that appear so essential for modern living is to improve our understanding of the physical, chemical and biological properties of natural systems and of how such properties are disrupted by exposure to chemicals.

In this final chapter, we discuss some of the steps that have been taken towards reaching the two goals outlined above. We analyse the current state of the field by focusing on several key issues presently under debate. We discuss the direction in which the field appears to be heading and suggest strategies that may improve our chances of ultimately reaching these goals.

6.1 Where have we been?

In 1978, Cairns and colleagues concluded that environmental scientists lacked a general understanding of how physical, chemical and biological phenomena interact to control the distribution of pollutants in nature. Since that time, vast amounts of data have been collected with regard to the influence of intrinsic physicochemical properties, external environmental factors and biological variables on pollutant fate, kinetics and toxicity. These data have been summarized in various review papers and in recently published textbooks (e.g., Rand and Petrocelli, 1985; Amdur *et al.*, 1991). Here we consider how the co-ordination of national and international efforts in the past has drawn attention to the environmental hazards of chemicals, and we provide a brief overview of the kinds of factors that can influence ecotoxicological phenomena.

6.1.1 Development of national and international organizations

Increasing public concern over chemical hazards, stemming in part from a number of tragic and widely publicized incidents, played a large part in ensuring the enactment of legislation designed to restrict the manufacture, use and release of chemical substances into the environment. In addition to public pressure, the desire to avoid trade barriers provided the primary motivation behind the standardization and coordination of chemical testing and control efforts (Lee, 1985). The Toxic Substances Control Act, enacted in the USA in 1976, helped to stimulate the development and standardization of chemical testing procedures. The Organization for Economic Cooperation and Development's chemicals testing programme was apparently launched in response to concern about the appropriateness of the Environmental Protection Agency's implementation strategy for the Act (Lee, 1985). The OECD Chemicals Group was formed to achieve 'the international harmonization of the philosophies of the approach to, the principles of and the methodology for assessing the potential hazard of chemicals to man and the environment' (Lee, 1985). To this end, members devoted a great deal of effort toward formalizing a set of guidelines for testing of chemicals which were 'not designed to serve as rigid test protocols' but which should be consulted 'whenever testing of chemicals is contemplated' (Organization for Economic Cooperation and Development, 1981). These guidelines were designed to 'provide a common basis for the acceptance of data internationally, together with the opportunity to reduce direct and indirect costs to governments and industry associated with testing and assessment of chemicals.' Data generated experimentally according to the test guidelines, and conforming to agreed principles of good laboratory practice, must be accepted among OECD countries as affirmed by the OECD Council Decision on Mutual Acceptance of Data (Organization for Economic Cooperation and Development, 1981). In addition to organizations such as the

OECD, various national (e.g., American Society for Testing and Materials) and international (e.g., International Standardization Organization) standards organizations have been important forces in the development of testing protocols and have focused political attention on the need to strictly control the manufacture, use and release of chemicals. While this type of activity has certainly had beneficial effects, there is always the danger that guidelines will be used as rigid protocols in practice. This use would clearly be contrary to the original spirit (Organization for Economic Cooperation and Development, 1981).

Despite increasing amounts of public, legislative and scientific attention drawn to the threat of chemical pollutants, progress in effectively dealing with pollution has been too slow to keep up with the chemical demands of society. Public and political concerns have been idiosyncratic and inconsistent, legislative decisions continue to be based on too little or inappropriate data, and scientific knowledge of pollutant fate and effects on natural systems is insufficient in many respects. The good news is that serious efforts to improve this situation continue.

6.1.2 Fate and effects: role of intrinsic chemical properties

A toxic substance's intrinsic physicochemical properties affect its fate and behaviour within organisms and ecosystems (Menzer, 1991). Properties such as molecular structure, water solubility and vapour pressure determine the rates and pathways by which chemicals move among environmental and biological compartments. The structure of a chemical determines its stability and persistence in the environment. For example, hormonally based insecticides break down rapidly in aquatic habitats, organophosphate and carbamate insecticides persist for intermediate lengths of time, and organochlorine insecticides tend to be highly persistent (Nimmo, 1985). Chemical structure also regulates the mechanism and degree of toxicity. For example, heavy metals tend to have very specific target sites and modes of action, whereas some organic compounds appear to exert toxic effects through a generalized narcotic effect (Lipnick, 1985).

The lipid solubility of an organic chemical, which can be described by the partition coefficient between octanol and water, is often correlated with the degree of bioaccumulation of the substance. Lipophilic substances can more readily cross cell membranes and tend to be more readily absorbed into the blood (Sipes and Gandolfi, 1991). For certain non-electrolyte organic compounds acting by a nonspecific mechanism, the octanol–water partition coefficient appears to correlate with acute toxicity in fish (Figure 6.1; see discussion of QSARs in section 6.2.3).

The hydrophobicity and vapour pressure of a chemical are the primary physical properties determining how it will be partitioned among particulate, air and water phases. The hydrophobicity of a chemical describes its

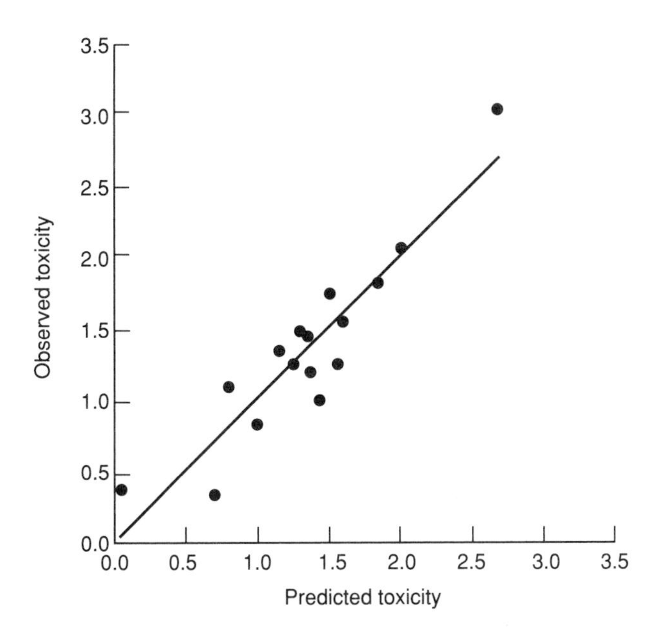

Figure 6.1 Relationship between observed acute toxicity (96-hour LC_{50}) of *Pimephales promelas* to 16 organic chemicals and toxicity predicted by the octanol–water partition coefficient of the chemicals. Redrawn from Nendza and Klein (1990).

tendency to migrate to nonpolar media in the environment (Swann *et al.*, 1983). In addition to influencing the bioconcentration potential of chemicals, water solubility determines the degree to which substances are distributed in aquatic systems. Chemicals having low water solubilities and which adsorb to particles will tend to become concentrated in soils and sediments, where they may be ingested by deposit-feeding or detritivorous organisms and possibly provide a future source of pollution.

In recent years, the basis for deciding which physical properties of chemicals are most relevant for predicting their environmental fate and effects has been strengthened. However, much still needs to be learned in order to synthesize these parameters into coherent and realistic models that can be applied to natural systems. This fact is driven home with particular force when we note that present QSAR models are statistical, rather than theoretical. Thus, even at the level of intrinsic physical and chemical properties, we are currently unable to develop good, predictive theoretical models derived from first principles.

6.1.3 Fate and effects: role of environmental variables

The ultimate sinks for most anthropogenic chemicals are the soil and sedimentary environments. Soil erosion is considered to be the principal carrier of pollutants found in water and is by volume the greatest single source of pollution to surface waters (Menzer, 1991). A number of external chemical and physical variables can alter the dynamics of chemical behaviour and the partitioning between water and soil (or sediment). Soil or sediment particle size and organic content influence the relative amount of chemical that is sorbed to particle surfaces. Temperature, salinity, oxygen concentration and pH have all been implicated as factors determining chemical behaviour and toxicity (Hahne and Kroontje, 1973; Menzer, 1991). These important physical and chemical factors interact to control toxicity primarily through their influence on a pollutant's bioreactive or bioavailable concentration.

With regard to environmental influences, perhaps most is known about metals. A great deal of study has addressed the physical and chemical factors regulating the speciation and bioavailability of metals, particularly in aquatic systems. Although the importance of controlling such factors in test systems has been widely recognized, this knowledge has been slow to find its way into the regulatory process.

Metal speciation has been recognized as perhaps the most critical determinant of metal uptake and toxicity. Free metal ions are usually the most toxic, probably because they are more readily assimilated by organisms than are metal complexes (Leland and Kuwabara, 1985). Chemical processes controlling speciation include inorganic complexation, chelation, precipitation and adsorption. Chelation is the formation of a metal-ion complex in which the metal ion is associated with a charged or uncharged electron donor, or ligand. Chelators tend to be nonspecific in their affinity for metals. They can enhance metal excretion as well as reduce uptake (Goyer, 1991).

Metal–metal interactions can have a large influence on uptake and toxicity (Goyer, 1991). For example, exposure to selenium via food enhanced the uptake of cadmium by the gills of the shore crab, *Carcinus maenas* (Bjerregaard, 1985). Also, exposure to selenite ($Se–SeO_3^{2-}$) in seawater resulted in increased levels of cadmium in the haemolymph and decreased the rate of cadmium excretion (Bjerregaard, 1988). Sundelin (1984) concluded that the toxicity of cadmium is decreased when the amphipod, *Pontoporeia affinis*, is also exposed to lead, even though cadmium and lead showed a mutually enhanced uptake when amphipods were exposed to a mixture of the two metals. Sundelin (1984) briefly reviewed the literature regarding the effects of metal–metal interactions on accumulation and toxicity. No coherent pattern was apparent. Synergistic, additive and antagonistic effects of metal mixtures have all been found. All that can be said is that observed effects depend on the specific combination and concentration of metals tested, the duration of exposure and on the species used.

Klerks (1987) noted that uncertainty in our estimates of bioavailability can confound interpretations of tolerance acquisition. Determining the actual concentration of pollutants to which organisms are effectively exposed under various environmental conditions has complicated past interpretations of pollutant toxicity and remains an important area for future research. Until the basic physicochemical processes determining bioavailability are worked out for a variety of systems, general patterns of response will be difficult or impossible to determine (National Oceanic and Atmospheric Administration, 1991).

6.1.4 Fate and effects: role of biological variables

Differences in accumulation and toxicity of chemicals among species can result from differences in morphology (for example, a shell or thick, chitinous integument may offer more protection than skin or scales), body size and shape, mobility (more mobile species may avoid exposure, for example), habitat and trophic level and probably many other factors. Differences in accumulation and toxicity occurring within species or within populations can be related to differences in sex, age, genotype and individual history (nutritional, exposure etc.). Often the youngest age classes are the most sensitive to the effects of pollution, but this is not always so (e.g., Sipes and Gandolfi, 1991; T. Forbes and Lopez, 1990).

Although food availability can be considered as an environmental variable, it influences toxicity largely through its effect on an organism's nutritional status. Thus, in the following discussion we include studies of the effects of food type and availability on toxicity under the heading of biological variables.

Starvation and nutritional deficiencies can cause substantial changes in the sensitivity of an organism to a toxicant. Often toxicity increases with starvation or food deficiency, but sometimes the reverse is true (Uthe *et al.*, 1980). Changes in food quantity and nutritional quality were found to have the largest effect on juvenile production in *Daphnia* spp. when compared to effects of light intensity, photoperiod and temperature (Lee *et al.*, 1986). At low food levels, effects of the other environmental variables became more important. Also, vitamin-enriched algae increased the number of juveniles produced, but only at high algal concentrations. At low concentrations, the authors found that vitamin-enrichment actually seemed to reduce juvenile number.

In some species, juveniles may be better able to tolerate environmental and nutritional stresses than adults. Adults of the deposit-feeding polychaete, *Capitella* sp. 1, appear to be more sensitive to reductions in environmental oxygen concentration and, to a lesser extent, food concentration, than are juveniles of this species. The interactions influencing the responses of worms to changes in environmental pO_2 and food concentration are complex and suggest that a change in the relative importance of these two variables occurs during development. The much more rapid starvation rates of adult worms, relative to juveniles, under hypoxic conditions is consistent with the rapid decline in reproductive output of established populations under conditions of food limitation (T. Forbes and Lopez, 1990).

Evidence from vertebrate toxicological studies suggests that trophic level and taxonomic affinity can frequently be uncoupled from toxic response

(Truhaut, 1977). For example, the weight-specific LD_{50} for penicillin is 300 times higher for the mouse than for the guinea pig. Triorthocresyl phosphate elicits an acute toxic response in hens and man, but not in rhesus monkeys, rats or rabbits. This cannot be due entirely to differences in metabolic pathway as at least the hen and the rat metabolize triorthocresyl phosphate in very similar ways. Tucker and Leitzke (1979) stated that:

> In a very general way, all species of animals show similar susceptibility to most chemical toxicants. Species differences become evident, however, as the degree of precision required by the toxicity test is increased.... The current empirical nature of the science of comparative toxicology demands that individual tests must be carried out for every species–chemical toxicity value needed.

This and more recent evidence suggests that broad generalizations based solely on taxonomic affinity are unlikely (e.g., Blanck *et al.*, 1984).

Some of the variability in the fate and effects of chemicals within organisms may be related to differences in organism morphology and physiology. For example, as ingested material (including toxic chemicals) enters an organism's gut a number of physical and chemical changes generally take place relative to the external medium. Whereas many organisms maintain an acidic gut to aid in the digestion of incoming food, others, especially some aquatic insects, maintain very alkaline guts with pHs up to 11 (Dadd, 1975). Redox conditions in animal guts typically range from slightly reducing to completely anoxic. As discussed in section 6.1.3, factors such as pH and oxidation–reduction potential can have a substantial influence on chemical behaviour and toxicity. Thus, species-specific differences in morphological (e.g., digestive) structure and function may help to explain the following.

1. why are some pollutants more rapidly accumulated than others?
2. why do species differ in pollutant uptake?
3. why does the relative contribution of food to total pollutant intake vary among species and pollutants (Decho and Luoma, 1991; Weston, 1990)?

With regard to the last item, Reinfelder and Fisher (1991) found that marine copepods retained the elements carbon, selenium, phosphorus, cadmium, silver, americium, zinc and sulphur with an efficiency that directly correlated with the presence of the elements in the cytoplasm of the copepods' diatom food. Elements associated with diatom cell walls and plasmalemmae were not retained. These results suggest that copepods obtain all of their nutrition from the algal cytosol and provide a likely explanation for the variability in absorption of different metals by these invertebrates.

6.1.5 Fate and effects: combined role of environmental and biological variables

Unfortunately, the ecotoxicological literature is too often devoid of testable hypotheses to explain or predict differences in chemical toxicity and behaviour. As a result, the published data frequently give one the impression of a miscellaneous, almost random, collection of minutiae. We suggest that one way to improve this state of affairs is to view important problems from an increasingly interdisciplinary perspective. We illustrated an example of this approach in the previous chapter when we discussed the influence of the resident biota on organic pollutant degradation. Here we outline how this perspective could be extended to deal with metal toxicity in sediments.

Although redox conditions are clearly an important factor regulating the migration and speciation of metals in aquatic sediments, different metals respond differently to changes in sedimentary redox potential. Studies of metal fate have been slow to yield generalizations concerning the role of environmental variables in controlling metal speciation and bioavailability. For example, Lu and Chen (1977) found that Fe and Mn were released from polluted surficial sediments to overlying seawater as conditions became progressively reducing. In contrast, cadmium, copper, nickel, lead and zinc were released in increasing amounts as conditions became more oxidizing, whereas release of mercury and chromium were not affected by redox conditions. As is often the case, no general explanation or hypothesis was put forward to explain these patterns. Fortunately the situation seems to be improving. For example, work by Di Toro and colleagues suggested that the bioavailability of metals may be controlled to a great degree by the amount of acid-volatile sulphide (AVS) present (Di Toro *et al.*, 1990). AVS is an empirically defined reactive pool of solid phase sulphide that is available to bind with metals. Removal of the metal ion from solution by sequestration as an insoluble sulphide has resulted in surprisingly low toxicities when metal concentrations were expressed on a sediment dry weight basis. Acute toxicity test results indicate that much better prediction can be achieved with normalization by AVS concentration instead of the usual sediment dry weight (Di Toro *et al.*, 1990). Because the pool of AVS maintained in a deposit is determined by the dynamic equilibrium between sulphate reduction and sulphide oxidation, it can be expected to vary both temporally and spatially. Interestingly, because the AVS pool is often positively related to organic loading, eutrophic environments may be able to suffer relatively greater inputs of metals without significant effect compared to less organically enriched environments. Thus one would expect clearly defined linkages and relationships between eutrophication, benthic ecosystem processes and metal toxicity.

Using an approach that could readily be extended to include the more commonly studied pollutant metals, Aller (1980) was able to accurately estimate Mn^{2+} and Fe^{2+} production rates as a function of depth within the sediment for Long Island Sound, USA, deposits using a one-dimensional transport-reaction model. We have illustrated this general type of approach in detail in Chapter 5 and have used it to generate testable hypotheses concerning the effects of key biological and physical variables on the distribution of organic pollutants in sediments. For metals, as with organics, the key variables needed for prediction of soluble (and therefore toxic) phases are particle density, sedimentation rate, porosity, particle mixing coefficient (for both biogenic and physical mixing) and a reaction-rate term. Several points are worth emphasizing. First, this approach can dramatically extend the usefulness of the relationship between AVS and acute metal toxicity by allowing general and broad-based predictions of bioavailability (pore water [Metal]) and toxicity in the field from a few important but relatively easily measured variables. Second, for metals as well as organics, fate and effect will be strongly linked through the effects of the biota on the sedimentary environment (Aller, 1980, Chapter 5). Third, note that the same key parameters are important for determining the fate and effects of both metals and organic pollutants (Chapter 5). This is an example of the level of generality we should be striving for and seeking to extend in all ecotoxicological research.

6.1.6 Mechanistic bases of pollutant effects

Pollutants can exert effects at all levels of biological organization. The ease with which we can detect effects tends to decrease with increasing system complexity. A wealth of toxicological literature is available regarding the mechanistic bases of chemical toxicity at the individual organism level and below (e.g. Amdur *et al.*, 1991). For compounds such as pesticides, the target sites and modes of action have been particularly well studied. Many hydrophobic organic chemicals appear to act through a nonspecific narcosis whereas many pesticides and most metals appear to have very specific targets (Leland and Kuwabara, 1985). It is generally believed that effects of chemicals at the molecular and cellular levels occur before physiological changes become evident, and likewise, that physiological changes provide an early warning system for effects at the population level. The belief that chemical effects at lower levels of biological organization, such as molecular and biochemical, can be linked to effects at higher levels of organization, such as population and community, and the idea that response time increases as one moves up the biological hierarchy, have provided the conceptual basis for the development of the biomarker approach (McCarthy and Shugart, 1990; Peakall, 1992). Although we have achieved substantial progress in our understanding of the mechanistic bases of chemical effects, the extent to which links can be established among effects at different levels of biological organization remains to be demonstrated.

Although the molecular, biochemical and physiological mechanisms through which pollutants exert their effects have been relatively well studied, much less is known regarding how effects at these levels are translated into effects on population dynamics and community structure and function. Thus, while certain biochemical or physiological changes induced by pollution have been assumed to provide an early warning of population or community-level effects, explicitly quantifying the connections between biochemical or physiological changes and changes at the population or community level (either theoretically or empirically) has been attempted by very few investigators (e.g., Koehn and Bayne, 1989; Calow and Sibly, 1990).

There may be fundamental reasons underpinning this lack of success to date. Ecologists are coming to the realization that a single-minded search for the underlying mechanisms, at the expense of studying the processes or systems of primary interest, is not the most efficient way forward. Peters (1991) has written that:

> The search for mechanism ... presumes that there will be some convenient and obvious stopping point, when the mechanism of one interaction is known and the researcher can presumably switch attention to another interaction. Experience suggests that this is improbable. The researcher is more likely to be sucked into the maelstrom of reductionism. [This is termed] the problem of infinite regress whereby each layer of mechanism is peeled back to reveal more and deeper sets of causal processes.

We suggest that the most efficient use of limited resources can be made by directly studying the systems of interest at the appropriate levels of complexity.

6.1.7 Genetics of pollutant resistance

There are few cases in which the genetic contribution to phenotypic variance in resistance under natural conditions has been quantified (e.g., Klerks and Levinton, 1989; Hoffmann and Parsons, 1991 for review). Studies of insecticide resistance suggest that the number of loci contributing to stress response traits is related to the mechanism of resistance and the stage at which it acts and by the intensity of selection. Resistance to specific stresses may have a very simple genetic basis which might involve, for example, a structural change in a target enzyme. Many cases of insecticide resistance, attributable to a single resistant gene, have been reported (Hoskins and Gordon, 1956; Hoffmann and Parsons, 1991). In contrast, stresses that have broad physiological effects, for example, reduced oxygen availability, may be expected to result in selection at a number of loci. Hoffmann and Parsons (1991) suggested that mechanisms of resistance causing exclusion of a toxicant are likely to have a simple genetic basis whereas resistance via changes in metabolic pathways may be more complex. There is evidence that extreme levels of stress, for example, pesticides applied at field doses to target insects, select for single genes having large phenotypic effects. Studies of less severe levels of stress (such as would be expected under chronic low-level pollution) have often found resistance to have a polygenic basis (reviewed by Hoffmann and Parsons, 1991).

The degree to which environmental variation modifies the phenotypic expression of a genotype (the width of a genotype's norm of reaction) is known as phenotypic plasticity (Hoffmann and Parsons, 1991). A phenotypically plastic genotype is expected to maintain relatively high levels of fitness under a variety of environmental conditions. However, there is some evidence that high levels of phenotypic plasticity are associated with low levels of stress resistance. Hoffmann and Parsons (1991) suggested that increased stress resistance may reduce the level of plasticity if the same mechanism controls the plastic response and genetic variation in stress resistance. They summarized the factors expected to influence the relative importance of plasticity versus genetic variation in controlling phenotypic responses to stress. Low plasticity levels in the population, high genetic variation for resistance or evasion, high costs of phenotypic plasticity, low costs of resistance or evasion, rare occurrence of periods of stress, rapid and severe environmental change and historical constraints would all be expected to favour genetic variation. The reverse conditions should favour stress responses involving phenotypic plasticity.

Although it appears that organisms already stressed by their environment (from either natural or artificial causes) are often less likely to tolerate exposure to a pollutant than are relatively unstressed individuals (Moriarty, 1983; Cairns and Pratt, 1986; Hoffmann and Parsons, 1991), certain species seem to be widely tolerant of a range of environmental perturbations. Positive correlations between resistance to different stresses have been found for stresses which exert similar physiological effects. However, resistance to stresses that represent opposite ends of a continuum, such as extreme heat and cold, may be mutually exclusive (reviewed by Hoffmann and Parsons, 1991). Whereas many tolerant species typically colonize disturbed or inhospitable habitats, not all colonizers are widely tolerant. For example, although *Capitella* sp. 1 is among the first species to recolonize organically polluted, hypoxic habitats (Grassle and Grassle, 1976), this species is not particularly tolerant of sulfide-rich (Warren, 1977) or low oxygen conditions (T. Forbes and Lopez, 1990). In addition to tolerance, factors such as fecundity, dispersal ability and reproductive mode play an important part in the success of colonizing species (Lewontin, 1965).

Luoma (1977) suggested that the demonstration of increased resistance by a population exposed to a pollutant provides evidence that the pollutant has caused biological effects, and suggests that other species in the community may also have been affected. Luoma (1977) argued that, to evolve tolerance, a pollutant must be in biologically available concentrations to limit the reproductive success of the less tolerant fraction of individuals in a population, in other words, there must be selection. Likewise, Klerks and Levinton (1989) suggested that screening for resistance might provide a valuable tool for identifying pollutant impact. Such a tool might be particularly useful for situations in which mixtures of potential pollutants are present and the goal is to identify which components in the mixture are having a toxic effect. Using this approach, only those toxicants to which increased resistance is demonstrated are indicated as having biological effects (assuming that cross-resistance is minimal, see Chapter 3). Blanck and others (1988) have extended this idea to the community level (see Chapter 3).

Several studies have suggested that the benefits of increased tolerance are not accrued without cost. Alleles determining insecticide resistance have often been associated with deleterious fitness effects when the insecticide is absent. However, the physiological cost of stress resistance is generally unknown and may be quite variable. Klerks and Levinton (1989) found that oligochaetes from a heavily cadmium-polluted habitat evolved resistance to this metal. After two generations of laboratory breeding, the offspring of the resistant worms were slightly less resistant than worms collected directly from the polluted site. The authors suggested that reduced tolerance could indicate an environmental component to the maintenance of resistance or a relaxation of the selection pressure when worms were cultured in clean sediment. If reduced tolerance were related to a relaxation of selection pressure, this would suggest that there is a cost to increased resistance and that this cost results in reduced fitness for tolerant worms grown under unpolluted conditions. Studies of plants adapted to contaminated soils have found that tolerant individuals are competitively inferior to nontolerant plants when grown on noncontaminated soil (summarized by Macnair, 1981). Investigations by Whitten *et al.* (1980, cited by Wood and Bishop, 1981) suggested that dieldrin-resistant genotypes of the sheep blowfly, *Lucilia cuprina* were at a disadvantage, relative to susceptible genotypes, in the absence of the insecticide. Thus, patterns of variation in stress resistance in space and time suggest that there are tradeoffs such that increased stress resistance may lead to decreased fitness under nonstressful conditions (Hoffmann and Parsons, 1991).

With the exception of pesticides, the study of the genetics and evolution of resistance to chemical pollutants has been largely neglected in ecotoxicological studies. In addition to adding insights into our understanding of the evolutionary process, study of the genetic basis of pollutant resistance has and will continue to provide an important tool for detecting and predicting pollutant effects on living systems (Luoma, 1977; Blanck *et al.*, 1988; Klerks and Levinton, 1989; Hoffmann and Parsons, 1991; Chapter 3).

6.2 Where are we now?

Cairns and Mount (1990) noted that over 9 million chemicals are listed in the Chemical Abstract Service's *Registry of Chemicals*, although only an estimated 76 000 are in daily use. In OECD countries, chemicals account for 10% of all trade, excluding products in which chemicals are critical components (Lee, 1985). Some of these substances are undoubtedly safe, either because of their inherent properties or because they are released in very small amounts. Others we know to be toxic, perhaps even very toxic. For the vast majority we have almost **no** idea of their capacity to disrupt or impair natural systems. Thus scientists and regulators are under increasing pressure to develop approaches that ensure adequate protection of our environment. The next few sections focus on how present knowledge is being applied to meet this challenge.

6.2.1 Efforts toward international standardization continue

As we discussed above, one of the primary goals of international organizations such as the OECD, has been the development and co-ordination of test protocols for the evaluation of potentially harmful chemicals. For example, within the section on 'Effects on Biotic Systems' (Organization for Economic Cooperation and Development, 1981) there are test guidelines for:

1. an alga growth inhibition test;
2. an acute immobilization and reproduction test for water fleas;
3. a fish acute mortality test;
4. a fish 14-day lethal and sublethal test;
5. an avian 5-day mortality test;
6. an avian reproduction test;
7. an earthworm acute mortality test;
8. a growth test for terrestrial plants;
9. a microbial aerobic respiration inhibition test with activated sludge.

In a summary of the report from the OECD expert group on ecotoxicology, which was assigned to formulate the test guidelines, it was stressed that 'the imperative prescription of one standard test species to be used in environmental testing is irrational. To do so would suggest an accuracy which does not exist and, moreover, would be contrary to the philosophy of model systems in ecotoxicological testing.' Despite these warnings, the European Community requires that notifiers provide a minimum pre-marketing set of data on the ecotoxicological hazard of chemicals consisting of acute toxicity data for a freshwater fish and *Daphnia* species along with some simple and very artificial degradation and bioaccumulation estimates (Commission of the European Community, 1989). Because of the stepwise nature followed in risk assessment protocols, results of this preliminary data set are critical for decisions as to whether more detailed tests should be carried out. Although the OECD guidelines were designed to provide only a general guide to performing tests and not rigid test protocols, notifiers completing pre-marketing data sets must indicate and justify any deviations from recommended protocols (Commission of the European Community, 1989).

The apparent ease with which recommendation can inadvertently become requirement is somewhat unsettling and suggests that a tighter collaboration between scientists and legislators is needed. Guidelines and regulations that are based on public or political pressure rather than sound scientific principles give society a false sense of security by suggesting that the problem of chemical toxicity is solved when, in fact, it is not. The unfortunate truth is that ecosystems continue to degrade despite increasing legislation and requirements. An obvious way to upgrade and maintain guideline development is to institute peer review of the proposed guidelines as is normally the case for a scientific proposal or research article. Lee (1985) criticized the development and adoption of guidelines at the international level as occurring with neither the benefits of the normal scientific peer review process nor with adequate testing by experienced laboratories. He attributed this situation largely to the slow development of standards and to the presence of political and public pressures. Political and public pressure, no matter how defensible, do not provide sufficient justification for the continuing lack of independent and rigorous scientific evaluation of environmental management strategies (guidelines and regulations).

It may be argued by some that the appointment of the various expert groups that periodically meet to discuss guideline development is sufficient in this regard (for example, the '*Daphnia* variability' and the 'extrapolation of effects' workshops, Chapters 4 and 6 respectively). It is not. There are several reasons why this strategy is inadequate. First, appointments to the various groups often tend to be arbitrary and can be politically rather than scientifically motivated. Second, the documents produced by the groups justifying the particular strategy taken are based on a consensus. The problem concerns the fact that there is no documentation of any vigorous discussion of the pros and cons of the important questions after the group decisions have been conceded. This effectively leads to a suppression of opposing views, a diffusion of accountability and a final document in which the range of viewpoints is not apparent or clearly defined. This makes the tradeoffs taken by the decision-making group unclear because there is no formal record of the opposing views. All too often decisions to promote or further develop new test systems or methodologies are taken by an expert group composed primarily of the developers themselves. This is unfortunate and to the detriment of decision makers and the public. Formal peer review would allow the written documentation of a wide range of scientific viewpoints on a given question. The final decisions could then be made explicitly in the context of the views aired in the reviews.

6.2.2 Debate on single-species tests continues

The debate concerning the relevance of single-species tests is hardly new. Issues concerning the appropriateness of laboratory-based, single-species toxicity tests for predicting ecotoxicological phenomena have been discussed at least since the early 1970s. In 1972, an amendment to the USA Clean Water Act prompted an evaluation of the ability of single-species toxicity tests to predict pollutant effects on ecosystems. Even then, the scientific community found the value of such tests to be low (Neuhold, 1986). At an American workshop held in 1977 in Michigan, participants concluded that determining the ability of laboratory-derived toxicity data to predict ecosystem effects was a topic of major concern in aquatic toxicology (Maki *et al.*, 1979). In a subsequent workshop held one year later to address this and other issues, Donald Mount (1979) raised the argument that, unquestionably, laboratory toxicity data can predict gross ecosystem effects, but that, unquestionably, they cannot predict refined effects. Rand and Petrocelli (1985) asserted that the results of single-species tests 'cannot be used to assess the chemical impact above this level of biological organization.' Neuhold (1986) expressed his criticism as follows,

> To project ecosystem effects from single-species laboratory toxicity studies is almost impossible. If toxicity information on all life stages of all species were available, and we understood the competitive and symbiotic relationships among individuals within a population and among species within the community, and we understood the relationships of environmental factors on each of the species and on the toxicant or toxicants, we could construct a model that could predict ecosystem level effects. Such a model, however, would be so large and so complicated that the feasibility of its construction and validation would be seriously questioned.

The bottom line is that both theoretical and empirical evidence strongly argues against the scientific soundness of using single-species toxicity tests to predict chemical effects at higher levels of biological organization. Interactions among species and between biotic and abiotic components of ecosystems are the primary factors preventing accurate predictions from single species to higher levels of organization. However, despite the above arguments and convincing cases presented by other scientists (e.g., Moriarty, 1983; Cairns, 1984; Barnthouse *et al.*, 1986; Blanck *et al.*, 1988; Baughman *et al.*, 1989; Underwood and Peterson, 1988; Lampert *et al.*, 1989) pointing out the strict limitations of single-species laboratory tests, a disproportionate amount of resources has been and continues to be devoted to fine-tuning and standardizing these tests. Some ecotoxicologists have actually suggested that first steps to improvement of the latest generation of distribution-based extrapolation methods should involve complementing the standard test battery with 'internationally accepted test guidelines on at least a mollusk and an insect species' (van Leeuwen, 1990). Denmark appears to be doing just that, albeit with a protozoan. The Danish Environmental Protection Agency is actively engaged in developing and promoting a new test for incorporation into the OECD guidelines using the ciliate *Tetrahymena.* It has been suggested that the inclusion of this organism will add a critical trophic link to the algae–*Daphnia*–fish suite of protocols. Given the highly questionable usefulness of single-species laboratory tests, one cannot help but wonder whether the addition of yet another single-species test will even marginally improve our ability to protect natural ecosystems from the threat of chemicals.

Part of the reason that single-species tests still prevail in ecotoxicology may stem from the fact that such tests have been used with success by toxicologists. The application of single-species tests in toxicology has been rational and effective because extrapolation of data is usually used to predict toxicity from one or a few species to only one species (humans) and, due to a variety of ethical and legal issues, toxicity testing of humans is not an option. Thus, toxicity tests using the rat, rabbit and other mammals have been invaluable for assessing potential human health hazards. Though mammalian test systems (particularly those for predicting carcinogenicity) have recently come under fire (addressed in letters to *Science*, vols **251**, **252**, 1991), it is certain that data available from human safety studies can provide valuable auxiliary information in environmental risk assessment.

Relative to current field testing schemes, it is often argued that laboratory toxicity tests are superior in terms of ease, speed and cost. These are some additional reasons that such tests have remained popular in ecotoxicology even in the face of heavy criticism for their lack of ecological relevance. In marked contrast to the high frequency with which the 'too expensive' argument has been raised against community-level or field tests, is an obvious absence of the detailed figures that would make such an argument convincing. We strongly suspect that, although this argument has become acceptable, it is often not justifiable (e.g., see Cairns, 1984 and Cairns *et al.*, 1986). This is especially true if the likelihood and cost of making a bad decision are taken into account. If we consider all of the effort and expense that are currently being devoted to validating and extrapolating single-species acute laboratory tests, the cost of developing such tests, relative to developing community-level and field tests, increases dramatically.

The options are limited. One can extrapolate from simple laboratory tests to the field or perform more complex tests. The current state of knowledge is such that in order to develop workable extrapolation methods for simple laboratory test schemes, our current understanding of how pollutants affect ecosystems must be improved considerably by more field-oriented research.

In their report summary preceding specific ecotoxicological test guidelines, the OECD expert group on ecotoxicology summarized the status of single-species tests (Organization for Economic Cooperation and Development, 1981).

> In ecotoxicology the response of populations and communities to stress is more important than the responses of individuals. Since, however, chronic studies on multi-species systems under natural conditions have not yet become available, a practical approach requires a stepwise procedure starting with single-species tests under laboratory conditions and establishing concentration/effect relationships within definite time limits.

It is clear from the above statement that single-species tests were never intended to fill more than a temporary position in the ecotoxicological evaluation of chemicals. But rather than altering this approach, there is a tendency on the part of some to recommend the addition of further single species tests. The inherent variability in community-level and field test systems has discouraged their development for regulatory use even though they may more accurately reflect the fate and effects of pollutants in natural systems. Although the use of single-species tests is valid for certain very limited purposes, such as determining the toxicity of a substance to a specified test population, those with a vested interest in their perpetuation as inexpensive and easy ecotoxicological tools sometimes go to great lengths to justify their continued use. For example, Alexander and Quick (1985) justified the use of single-species tests by industry because 'there is a very large data base and because test methods and test species have been standardized. Also the interpretation of results is well understood by toxicologists and regulators, test reproducibility is good, and there is now considerable regulatory recognition and dependence upon these methods.' Historical precedence can never be sufficient justification for continuing to use inappropriate tools.

Many publications have been written in an attempt to sell the single species approach as an effective short-cut for more complicated testing schemes. However, there has yet to be proposed a convincing argument as to why we should expect the acutely toxic response of a single species to be in any way consistently related to the ecological response of the same or other species, not to mention intact communities, in nature. At a 1988 OECD workshop on ecological effects assessment, a working group for the aquatic environment concluded, 'In light of the complexity of the aquatic ecosystem, no scientific basis for extrapolation of test data exists, be it from species to species, from species to communities or ecosystems, or from acute to chronic' (Organization for Economic Cooperation and Development, 1989). Although practical considerations may require that we continue to rely on single-species tests for the time being, the most rational long-term environmental management strategy is the development of testing strategies that more accurately assess the effects of chemicals on functioning ecosystems. For now, we are doing about as well as one could hope, given our ignorance. Truly new and effective tools must await results from the implementation of research programmes aimed at increasing our fundamental understanding of the effects of pollutants on populations, communities and ecosystems.

6.2.3 Physicochemical predictors of pollutant toxicity and fate

In an attempt to cope with the large number and variety of chemicals released into the environment, great efforts have been directed toward developing predictive approaches for assessing pollutant fate and effect. Quantitative structure–activity relationships (QSARs) are used to predict the biological activity (toxicity, bioaccumulation, biodegradation) of chemicals on the basis of their physicochemical properties. QSARs are statistical relations in which the toxicities (or behaviour) of compounds are related to one or more structural parameters via uni- or multivariate linear or nonlinear regression. QSARs have been most successfully applied to predicting the acute toxicity of non-reactive, non-ionized organic chemicals to aquatic organisms (Deneer and Hermens, 1988). Such chemicals tend to exert their toxicity via a nonspecific narcosis (anaesthesia). Because it is believed that the minimum toxicity of a chemical is limited by its hydrophobicity, narcosis is often referred to as baseline toxicity. For chemicals that interact with specific reaction sites, QSARs based on hydrophobicity alone may underestimate the true toxicity. QSARs have been judged effective for describing chronic toxicity (Deneer, 1988), bioaccumulation and the toxic effects of mixtures of chemicals (Calamari and Vighi, 1988). Most work in this area has focused on aquatic organisms, and it has been argued that successful application of the approach to terrestrial systems requires that inherent differences in routes of exposure be considered (Calamari and Vighi, 1988).

In order for QSARs to be useful for predicting toxicity, it is necessary that the mode of action through which a chemical exerts its toxicity be determined. This is because equations based on one or two physicochemical properties appear to adequately predict toxicity only for groups of chemicals sharing a common mode of action. Only rarely can the mode of action be predicted *a priori* from the structure of a chemical (Deneer and Hermens, 1988). Deneer (1988) recommended that efforts be focused on developing methods that will allow such predictions to be made, particularly for selectively acting compounds.

The mode of toxic action cannot be predicted from chemical structure alone because the properties of the target organism, for example, the metabolic pathway that the chemical takes once inside the organism, are also important. Even when the metabolic pathway of a chemical is known, surprising and unpredictable differences in toxicity have been found among species (Truhaut, 1977). Thus, although the physicochemical properties of a substance may be directly related to its molecular structure, additional and sometimes unknown factors influence a substance's biological activity and will prevent general predictions of toxicity, bioaccumulation etc. to be made on the basis of structure alone (Moriarty, 1984). Given that interspecific variability in toxicity for single compounds can span several orders of magnitude (Table 6.1) (e.g., Blanck *et al.*, 1984) and that differences in toxicity for the same test population and compound have occurred up to at least a factor of five (Sprague, 1985), we should not expect QSAR estimates of toxicity to replace more direct measures of toxicity in risk assessment and environmental management strategies.

To date, quantitative risk assessments have been based primarily on toxicity data with less emphasis placed on biological breakdown and remineralization of pollutants. Efforts to include biodegradation in environmental risk evaluations have tended to be either site specific or wholly qualitative (Shimp *et al.*, 1990). Biodegradation rates can be estimated using radiolabelled materials or specific analytical techniques (e.g., High Pressure Liquid Chromatography (HPLC), Gas Chromatography/Mass Spectrometry

Table 6.1. Median and range of EC_{50} values obtained for 19 chemical compounds tested on 13 algal species.

Chemical	EC_{100}(mg l^{-1}) median	EC_{100}(mg l^{-1}) range
Alamine 336	4.0	1.0–8.0
Aliquat 336	0.5	0.25–1.0
Copper sulphate	0.5	0.13–1.0
Copper sulphate + EDTA	50	13–200
DCMU	0.13	0.13–8.0
Disodium hydrogen arsenate	400	3.1–≥6400
Glyphosate	11	2.8–23
Hydroquinone	20	0.31–80
Mercuric chloride	0.50	0.063–0.50
Paraquat	0.025	0.0063–0.40
Phenylmercury acetate	0.063	0.0020–0.25
Potassium dichromate	1.0	0.13–16
Primene JM-T	1.0	1.0–≥8.0
SAN 9789	0.100	0.025–≥6.4
Sodium lauryl sulfate	500	31–4000
Sodium nitrite	4600	580–9200
Wastewater	20	10–≥80
Tributyl phosphate	50	25–≥100
Tributyltin chloride	0.13	0.063–2.0

Data from Blanck *et al.* (1984).
EDTA = (ethylenedinitrilo) tetraacetic acid; DCMU = 3- (3,4 - Dichlorophenyl - 1,1-dimethylurea)

(GC/MS)). The former approach has the advantage of simplifying the detection of compound or its breakdown products in various complex environmental compartments, whereas the latter allows breakdown products to be directly measured (Shimp *et al.*, 1990). Determining the importance of biodegradation in the removal of chemicals from environmental compartments depends not only on the biodegradation rate (as a function of temperature, oxygen etc.), but also on the residence time of the chemical in the compartments of interest. Any criterion for 'acceptable' biodegradability based on half-lives alone may underestimate, or overestimate, the importance of biodegradation in a particular environment (Shimp *et al.*, 1990).

As many chemicals entering aquatic environments may be rapidly lost from the water column by adsorption and transport to bottom sediments, it is important that realistic biodegradation models are developed for freshwater, estuarine and marine sediments. Biodegradation processes in estuaries are particularly poorly understood. Unfortunately, predicting chemical residence times in sediments is very difficult and is complicated by the fact that a number of important controlling variables, such as redox potential, infaunal community structure, vary with time and depth. Thus some authors have explicitly chosen to exclude sedimentary compartments in the development of practical biodegradation models (Shimp *et al.*, 1990). This is a major gap in our present understanding which needs to receive greater research priority in the future.

6.2.4 Need for less primitive test systems

We have discussed the limitations of laboratory toxicity test systems elsewhere (section 6.2.2). In the following paragraphs we outline analogous problems associated with the standardized tests currently available to estimate chemical degradation rates.

The OECD *Guidelines for Testing of Chemicals* organize biodegradability tests into three levels of increasing complexity: **ready** biodegradability tests, **inherent** biodegradability tests and **simulation** tests. Tests of ready biodegradability are believed to provide worst-case scenarios by providing conditions which permit only limited biodegradation to occur. Inherent biodegradability tests are designed to favour biodegradation by, for example, allowing prolonged exposure to microbes. Simulation tests are, in theory, designed to measure biodegradation rate 'under some environmentally relevant conditions'. The guidelines provide five standard tests for ready biodegradability in water, three tests for inherent biodegradability in water, one test for inherent biodegradability in soil and one simulation test which is intended to determine 'the ultimate biodegradability of test materials under conditions which simulate treatment in an activated sludge plant'. There is little reason to expect such tests to predict biodegradation rates in nature to within several orders of magnitude. In fact, for marine sedimentary systems, we know that they do not (Chapter 5). Also of importance is the fact that there is no presently accepted general guideline for sediments despite the fact that sediments are major sinks for pollutants.

As we argued in detail in Chapter 5, factors that we expect to exert important controls on pollutant fate, such as soil or sediment structure and the abundance and activities of macrofaunal organisms, are completely neglected in most test systems. To our knowledge no attempts have been made to examine the potential importance of feedback loops between biotic and abiotic processes in controlling pollutant fate and effects in nature. The tools to address such questions are readily available. For example, geochemical models, based on reaction–diffusion theory, have now been developed to the degree that we can begin to predict the depth profiles and inventories of certain chemical species as a function of reaction rates, animal body size and animal density (Berner, 1980; Aller, 1980; 1988). The approach is extremely powerful and has the valuable feature of explicitly linking chemical fate with biological effects. These diagenetic models have been shown to predict processes such as sulphate reduction and denitrification quite well in the field and represent precisely the type of approach that is needed to begin to deal more directly with the complexity of natural systems.

6.2.5 Need for more theory

As Box (1976) has so aptly put it, 'Science is a means whereby learning is achieved, not by mere theoretical speculation on the one hand, nor by the undirected accumulation of practical facts on the other, but rather by a motivated *iteration* between theory and practice'. The published literature suggests that ecotoxicologists have directed the bulk of their efforts toward amassing empirical data on the behaviour and effects of specific pollutants on selected species and have placed less emphasis on developing conceptual frameworks with which to generalize the analysis of pollutant-caused perturbations of the natural world. Models that have been applied to ecotoxicological problems have tended to be descriptive, rather than theoretical. One example concerns the class of predictive models of chemical toxicity, bioaccumulation and biodegradation based on quantitative structure–activity relationships (QSARs). Although statistical models are often helpful over the short term, the lack of understanding implicit in their use can cause problems when we try to extend our knowledge. Because we do not understand why models based only on statistical methodologies work, we are often at a loss about how to proceed when predictability breaks down.

Ecosystem theory, for example, stability–diversity, production and cycling of matter or energy, has been accused of being of little use in impact assessment. Suter (1981) condemned ecosystem theory for lacking credibility and predictive capability. He argued that theories are promoted on the basis of weak evidence and that their predictions have neither the necessary specificity nor absoluteness required for successful impact assessment. Without doubt, current ecosystem theory needs substantial improvement before it becomes an important practical tool. Yet improving the state of ecosystem theory so that it can be more effectively applied to practical problems, such as impact assessment, will require that environmental managers be willing to take a more far-sighted view that supports long-term research into basic ecological processes. This must be done because, 'despite the many flaws of existing ecological theory, there is no escaping the fact that the phenomena of interest in ecological risk assessment operate on scales of space, time and complexity that cannot be captured in laboratory test systems' (Barnthouse *et al.*, 1986). Based on our own experience, it is a major challenge to convince regulatory agencies that such a long-sighted view will pay off. In consequence, most current resources are being focused on fine-tuning old testing methods in attempts to solve what are often trivial problems.

6.3 Where are we going?

6.3.1 Validation and extrapolation in ecotoxicology

'Validation' has become a buzzword in ecotoxicology. But like many such buzzwords, it has suffered from inappropriate and careless use. Yet the performance of 'reality checks' of ecotoxicological models, particularly for tests involving some form of extrapolation to higher levels of organization, has properly become an area of increasing concern to those involved in developing and standardizing protocols.

Caswell (1976) provided a thoughtful and articulate discussion of the validation process for predictive and theoretical models. He noted that for predictive models (such as QSARs) the truth or reality of the model is never at issue. Here the major concern is to evaluate the model's accuracy and the range of conditions over which it is useful. Criteria for determining *a priori* whether a predictive model is validated may include the prediction error, proportion of explained variance and various cost–benefit factors. In addition, sensitivity analyses can be useful for determining where model refinement will have the greatest payoff. Although validation of the assumptions of predictive models may be useful for improving the model's predictability, validating the outcome of the model is of primary importance. Again, this is in contrast to the development of theoretical models such as the community tolerance distribution (discussed below), for which tests of the underlying assumptions are a critical step in corroborating or refuting model predictions.

A fundamental unanswered question in ecotoxicology concerns the extent to which ecosystem-level effects of pollutants can be understood or predicted from tests at lower levels of biological organization. Much research attention during the 1970s and '80s has been directed toward developing new and better test methods and identifying ideal test species, indicator organisms and biomarkers. Such developments contribute to the continuing evaluation and improvement of methods for assessment and prediction of pollutant effects.

Given the impracticality of testing all species for their sensitivity to chemicals and other pollutants, regulators and managers have generally relied on toxicity data gathered for a few selected species. In contrast to information obtained in human toxicological studies where data for a few surrogate species are extrapolated to humans, ecotoxicological testing requires extrapolation from a minimal number of test species to a vast number of species varying in taxonomy, size, life history, physiology and geographical range (Cairns and Mount, 1990).

Given that current environmental regulations rely very heavily on single-species acute toxicity tests for evaluating the potential effects of chemical pollutants in natural ecosystems, great concern has arisen over whether the results of such tests can predict effects in the field. Several methods have recently been advocated for evaluating the effects of chemicals on all species in a community from single-species toxicity tests (Stephan *et al.*, 1985; Kooijman, 1987; Van Straalen and Denneman, 1989; Aldenberg and Slob, 1991; Wagner and Løkke, 1991; but see Forbes and Forbes, 1993). These methods differ primarily in the assumed underlying distribution curve of species sensitivities and the fraction of species for which protection is sought. If one intends to use these methods for the protection of ecosystems then they all assume that a community consists of non-interacting species and consider community effects to arise solely from the sensitivity of member species to a toxicant. These extrapolation procedures are based on laboratory toxicity data for very few (from three to eleven) species. Because these models represent very important first attempts at projections from lower to higher levels of biological order and deal with interspecific variability in toxicity, we examine them below in some detail.

Wagner and Løkke (1991) emphasized the need to select test species that are representative of the ecosystem because, they argued, if the species represent a small part of the distribution curve, the variance among species will be underestimated and the protection concentration for the toxicant in question will be too high. The precise meaning of 'representative' is left unclear. In practice, 'representatives' are deliberately selected from different taxa and trophic levels. Wagner and Løkke suggested, 'A large variation of endpoints and taxa in combination with a limited number of test species may cause serious errors in the extrapolation procedure because of wrong assumptions of the distribution model or the lack of representativeness of the selected taxa.' Attempts to select representative species could result in the construction of a toxicity distribution that is more uniform than normal and thus overestimate the variance. Selecting a very sensitive species, a moderately sensitive species and a tolerant species to encompass the range of possible toxicity values for a community will overestimate the true community variance because the extreme values are sampled in greater proportion than they actually exist. The goal should be to design sampling schemes in which *random* species samples, rather than *representative* samples, are selected. Unfortunately, the selection of so few species as input into these extrapolation models (3–11) provides very little power for tests of the shape of the resulting distribution.

As discussed elsewhere in this volume, the usefulness of laboratory toxicity tests such as LC_{50} is quite limited if the goal of testing is to predict effects of pollutants on ecosystems. Comparison of the responses of different members of a community to pollution exposure can provide insight into whether and to what extent the community response can be explained as a result of differential species tolerances versus changes in community structure, competition or other indirect effects (Baker and Crapp, 1974). Given an appropriate experimental design, it should be possible to partition the variance in community response into additive components, for example, single species toxicity values, and nonadditive, such as species × environment interactions. Because there is no shortage of pollution incidents available for observation and study, it could be quite useful to attempt to link observations of community-level responses to pollution with acute toxicity test data for community member species. Few researchers have yet taken such an approach. However, it is one we feel would go a long way towards resolving the controversy surrounding the usefulness of single-species laboratory tests.

Maximum acceptable toxicant concentrations (MATCs), which are often used in evaluating chemical risk to ecosystems, are typically based on a number of assumptions related to the biological variability of toxicity. These assumptions are a critical and untested feature of the current batch of extrapolation models for what is optimistically termed 'refined effects assessment' (Organization for Economic Cooperation and Development, 1991). The most recent models have evolved from earlier 'assessment', 'application' or 'safety' factor approaches which attempted to estimate toxicant concentrations above which one would expect adverse effects on ecosystems. In this context we define an assessment factor as simply a number that is applied to single-species toxicity data to adjust the effects concentrations (L(E)C_{50}, NOEC etc.) to estimate a maximum acceptable toxicant concentration for a community or ecosystem (Organization for Economic Cooperation and Development, 1991). The recently developed extrapolation methods have been proposed to quantify a limited type of community response to a toxicant. If the assumptions hold, one can define an exposure level at which, say, only 1% of the community would be adversely affected. For example, one could estimate a concentration such that the NOEC for only 1% of the species in the ecosystem would be exceeded.

Proponents of the extrapolation methods have claimed that these methods can be 'considered to be an acceptable approach for the protection of the structure and function of aquatic ecosystems' (Van Leeuwen *et al.*, 1992). Obviously, the usefulness of these extrapolation models hinges critically on the validity of the assumptions. The core assumptions employed by the more sophisticated distribution-based extrapolation techniques are the following (Organization for Economic Cooperation and Development, 1991).

1. The distribution of species sensitivities in natural ecosystems closely approximates the postulated theoretical distribution. So far the log-triangular, log-logistic and the log-normal have all been hypothesized.
2. The sensitivities of species used in laboratory tests provide an unbiased measure of the variance and mean of the sensitivity distribution of species in natural communities. The variance and mean of the laboratory data are the parameters used to estimate the hypothetical natural sensitivity distribution. This means that the species chosen and tested to estimate the desired sensitivity distribution must be a random sample from that distribution.

If we continue further and wish to consider the ecological relevance of the methods for accurately assessing the hazards of chemicals to natural communities and ecosystems we must also make the following additional assumptions.

3. By protecting species composition, community function is also protected. We know, however, that community structure and function are often uncoupled (e.g., Schindler, 1987; Gray, 1989).
4. We must also assume that interactions among community and ecosystem members can be ignored. Again, this assumption may often be invalid. For example, reductions in the abundance of influential species (such as keystone species; Paine, 1974) will have drastic effects on the abundance of

many other species in the community. The extent to which influential species are either more or less susceptible to the effects of pollutants is unknown.

We discuss below some problems related to the use of these models in the context of the current OECD work. We have focused on the OECD report in some detail because of its central importance in regulatory decision making.

Before extrapolation models can be used with any confidence, they must be evaluated, or in the current jargon, validated. This entails testing the validity of assumptions 1 to 4 above and requires that the community or ecosystem of interest be operationally defined so that it can be sampled. Operational definitions of concepts such as **community** and **ecosystem** have eluded ecologists (and ecotoxicologists) in the past and have been the source of a number of problems that have seriously impeded hypothesis testing and theory development (Peters, 1991). Until the biological system of interest is defined precisely, it cannot be sampled properly and the validity of the extrapolation models cannot be determined. We are forced to define the ecosystem, community, or assemblage we wish to protect. Open acknowledgement of this fact places the problem in proper perspective. We can never know whether the hypothesis that the ecosystem sensitivity distribution follows, for example, the log-logistic distribution is valid until we know what constitutes the ecosystem. That we cannot make progress until we have developed operational concepts seems obvious, yet expert groups continue to dance evasively around the problem (e.g., Organization for Economic Cooperation and Development, 1991). Ultimately, the definition must involve political and cultural as well as scientific considerations because society (or its representatives) must consider the question of what qualifies for protection.

Thus the first step in what should be the evaluation before use of the current set of extrapolation models must be the development of an operational definition of the ecosystem or community because that is the focus of the modelling effort. We suggest that it is better to think of sensitivity distributions as applying to communities or, even more appropriately, assemblages, rather than ecosystems because of the complete lack of an abiotic component in their theoretical basis.

Oddly, the official OECD approach to the evaluation of extrapolation models appears to be relatively unconcerned with the validity of the underlying assumptions. A workshop comparing newly developed extrapolation methods was held in December, 1990 in Arlington, Virginia, USA (Organization for Economic Cooperation and Development, 1991) and actually recommended that, '**The use of the extrapolation methods for refined effects assessment** needs to be carried forward based on species sensitivity distributions, i.e. the log-triangular, the log-logistic, and the log-normal methods' (p. 38) [emphasis added] and, 'Because these extrapolation models have **reasonable** underlying assumptions, and have been **accepted for regulatory use**, there are good bases for their application in assessing the hazards of chemicals' (p. 21) [emphasis added].

Thus, the report concluded that if the assumptions **seem** reasonable (and we have started to use the methods anyway) we are justified in continuing to use the methods. But how do we know the assumptions **are** reasonable? The answer is that we do not. There are currently no acute or chronic toxicity data based on random samples from a well-defined ecosystem that can be compared with any of the hypothetical species sensitivity distributions. Despite claims to the contrary, we have not checked the validity of the distributional assumption with regard to natural ecosystems. For example, the workshop report cites the laboratory toxicity data used by Kooijman (1987) (which was originally compiled and reported by Sloof *et al.* (1983)), as data that was 'a random selection of species.' From what statistical population was it a random sample? The similarity of these data to those obtained from a few standard laboratory test species is then proposed by the report as a corroboration of the general distributional assumptions of the models (p. 37).

As part of the justification of the approach there has been a statistically misguided tendency to attempt to identify so-called 'representative' species. For the most part, these species are those already used in standardized laboratory toxicity tests. For example, the report states that, 'Variation among species is not a random error since selection procedures require a **diversity** of organisms in the data set' [emphasis added]. This is true; however, the report continues with, 'This diversity is intended to increase the likelihood of having both tolerant and sensitive species, **in order to improve the estimation** of the desired percentile' (p. 33) [emphasis added]. This is incorrect. The proper goal of these extrapolation studies should be to estimate the desired percentile as accurately and reliably as possible. This is best done by randomly sampling the statistical population of interest, which must be clearly defined before it can be sampled. Selection of a diversity of organisms that are somehow representative of the ecosystem will only act to bias the estimated mean and variance of the hypothetical distribution to some unknown degree.

It is unclear why the expert group should recommend that the use of these much more complicated and cumbersome extrapolation methods be 'carried forward' in preference to the previously used and much simpler assessment factors; factors that were also hypothesized to provide some degree of protection for the biota of concern. Previous approaches have recommended that these simple factors be applied to values from chronic or semi-chronic data sets for a given chemical. For example, dividing the lowest of three laboratory-determined NOECs by 10, where one value each has been obtained for a fish, a crustacean and an alga has often been used in the past for screening new and existing chemicals on the basis of very limited data (Organization for Economic Cooperation and Development, 1991).

Table 6.2 Comparison of 'extrapolation methods' based on laboratory NOEC values for eight selected compounds.

Compound	Lowest NOEC/ 10 (base set)	95% confidence value	Difference	% Difference
Potassium dichromate	0.01	0.010	0	0
Sodium bromide	1	0.21	-0.79	-79
TPBS	0.1	0.042	-0.058	-58
2,4 DCA	0.0032	0.0080	0.0048	150
p-NT	0.1	0.10	0	0
DNOC	0.01	0.0027	-0.073	-73
Dimethoate	0.0032	0.00036	-0.00284	-89
Pentachloro phenol	0.0032	0.00030	-0.0029	-91

Data from Organization for Economic Cooperation and Development, 1991.

To be approved for use by the OECD expert group, the distribution-based extrapolation methods should offer some unique or distinct advantages over the much simpler and cruder methods employed to date. Table 6.2 is a condensed version of one found in the Arlington workshop draft report on extrapolation (their Table 1). In the report the table was used to compare levels of concern calculated using chronic NOEC data determined for a limited set of organisms (an alga, a crustacean and a fish) as well as values for 11 (eight additional) species using both new and old extrapolation methods. The first data column of extrapolated values in Table 6.2 corresponds to the lowest NOEC from the set of three species divided by 10 and the second is what the OECD has termed the 'log -logistic distribution 95% confidence value' (here abbreviated 95CV). By definition, the 95CV is a value determined such that it will protect 95% of the species comprising the hypothetical ecosystem sensitivity distribution 95% of the time (Organization for Economic Cooperation and Development, 1991; Aldenberg and Slob, 1991; Wagner and Løkke, 1991).

We can now ask the following question. Are the newer and much more complex extrapolation methods an improvement over the previous simple and arbitrary methods? This was presumably one of the questions addressed at the workshop. We note that although the OECD draft report recommended carrying forward the use of these refined extrapolation methods (the **use**, not just the testing of assumptions), we could find no specific statements in the report justifying this conclusion. Because these methods have been recommended for use with three or more data values and because there is a general consensus among ecotoxicologists that chronic or semi-chronic NOECs are more relevant as a basis for extrapolation, we compared the predictions generated using the lowest NOEC from the set of three species divided by 10 and the 95CV method using the same data set employed by the OECD (Table 6.2).

Comparing the two columns we note that two predictions are the same regardless of the method used. Of the remaining six, the 95CV method predicts five values which range from approximately 60–90% lower. In the case of 2,4 DCA, the 95CV method predicts a value that is higher by 150% (Table 6.2).

At first it looks as if the 95CV method may tend to be a little more conservative and predict lower values in general. To test the hypothesis that the 95CV values are different from the assessment factor values, we compared the two groups of predictions using the nonparametric Wilcoxon signed ranks test for paired observations (two observations on each chemical) (Hollander and Wolfe, 1973). We used a nonparametric test because the variances of the two groups were extremely heterogeneous (s^2_{NOEC}:0.345, s^2_{95CV}:0.074; Bartlett's test, $df = 1$, $P = 0.0006$, s^2 = sample estimated variance). We also know from Monte Carlo simulations that estimated 95CV values, especially for small numbers of test species (five or less), deviate markedly from normality (see Aldenberg and Slob, 1991, their Figure 3).

Comparison of the paired results for each of the eight chemicals in columns 2 and 3 in Table 6.2 using the Wilcoxon signed ranks test provides very little evidence for a difference in the predictions of the two extrapolation methods ($P = 0.116$) (Hollander and Wolfe, 1973). We conclude that advocating the newer methods simply on the basis of their apparently more conservative predictions cannot be justified.

An additional theoretical problem with all extant extrapolation methodologies is the well-recognized lack of incorporation of higher level ecological processes (Organization for Economic Cooperation and Development, 1991, van Leeuwen, 1990). In their extension to the protection of ecosystems, all sensitivity distribution models must make two assumptions. First, they assume that there are no interactions among species. Second, one must assume that structural and functional changes are coupled, or alternatively, that functional changes can be ignored or are unimportant. With regard to interactions, one form mentioned above and commonly observed by ecologists concerns the concept of a 'keystone' species (e.g., Paine, 1974). Certain species are simply more important in influencing the structure and function of the communities to which they belong. If critical species happen to fall into the unprotected group, the impact of the toxicant on community structure and function will be way out of proportion to the fractional contribution of these species to the sensitivity distribution.

The degree to which structure and function can be decoupled during pollutant stress is currently a matter of debate. One school of thought maintains that detectable structural changes will occur before functional changes are manifest. The proponents of this line of reasoning suggest that feedback mechanisms at the population, community and ecosystem levels may often tend to preserve function even while significant structural changes are occurring (e.g., Schindler, 1987; Gray, 1989). At the other end of the spectrum are those who maintain that ecosystem functional changes in response to pollution are either likely to occur first or are more easily detected than structural changes (Crow and Taub, 1979). Theoretical considerations suggest that what we are able to observe as structural and functional components of ecosystems are not always simply related to each other (O'Neill *et al.*, 1986). This dual organization may arise from structural constraints that operate on organisms and functional constraints that operate on processes (see O'Neill *et al.*, 1986 for a more complete description). Exceptions to this general decoupling rule may occur when a single species dominates a function, that is, in communities that possess a 'functional keystone species' (after Paine, 1966). It seems reasonable to suggest that there are species whose removal would affect ecosystem processes. Finally, we would expect ecosystems with impoverished biota to show a much greater structure–function coupling when perturbed by pollutants. If there is little or no functional redundancy, then removal of a species may severely affect ecosystem processes. We suggest that much apparent inconsistency of results to date may be resolved by studies of functional redundancy and the degree of structural complexity of pollutant-stressed communities.

But what about the method of dividing the lowest of a set of NOEC data by some assessment factor? One important objection to this strategy is the following. As data for more species are accumulated for a particular chemical, the lowest value of the lot can only get lower (van Leeuwen, 1990). Thus as more toxicity information becomes available for a compound, the 'acceptable' concentrations must decrease. This situation is clearly unreasonable. A better approach would be to divide the median or geometric mean value of the available data set (base set or greater) by some arbitrary value. This estimator would be relatively robust to wildly different input values based on selections of 'representative' species. Choice of the median has the additional advantage of making no distributional assumptions. The method could be made as conservative as desired simply by adjusting the extrapolation or assessment factor. Although application of a constant factor to the median value is arbitrary it would have the desired effect of reducing the variance in the extrapolated values without the need for employing untested assumptions. This method is also less complex. If the predictions of two or more methodologies are indistinguishable, parsimony demands that we not use increasingly complex methods unless they have clearly demonstrated an increased usefulness. Mathematical treatment of the problem should be trimmed to the minimum necessary to address the problem without obscuring the main point. The more complex extrapolation methods have not provided improved predictive power over the cruder methods.

It has been suggested that the recently developed extrapolation models (e.g., that of Van Straalen and Denneman, 1989) are a 'useful tool in ecotoxicological effects assessment but should only be used if more than three (semi) chronic test data are available' (van Leeuwen, 1990). Specifically, the current test data should be supplemented by 'internationally accepted test guidelines on at least a mollusk and an insect species' (van Leeuwen, 1990). This amounts to putting the cart before the horse. First, as mentioned above, these methods have not been shown to be either more accurate or more conservative than the simpler assessment factor approaches. Second, and most important, focusing on improving the estimates generated by a model of questionable validity is misplaced effort. If we are truly serious about eventually applying these models for regulatory purposes, we should begin by testing the assumptions. Focusing effort and increasingly limited resources on the selection and international acceptance of two additional 'representative' species is inappropriate.

There is also a more insidious problem with the recommendation to use the 'refined' extrapolation methods described above. This has to due with the fact that by failing to rigorously test the assumptions of these new models before incorporating them into the ecotoxicological 'toolbox' we run a serious risk of deluding ourselves and the public that we have made substantial progress when in fact we have not. Just because these models have a more sophisticated outward appearance and seem to incorporate reasonable assumptions *a priori* does not justify their uncritical incorporation into regulatory protocols. Most ecotoxicologists justifiably abhor the use of arbitrary assessment factors, but the wish for improved chemical evaluation methods must not obscure critical scientific evaluation of their use. Thus we recommend that the older methods remain in use until the assumptions of these new models have been tested rigorously. Immediate improvements such as the use of a median or geometric mean rather than lowest observed NOEC value may be employed to improve the stability of the estimates in the face of increasing information. While unarguably crude, the less refined assessment factor approaches should be used as the most parsimonious current approach, a monument to our ignorance and a spur to rapid, rigorous and critical evaluation of the next generation of methods.

6.4 How do we get there from here?

6.4.1 Selecting the right tools

Among the impediments to successful assessment and management of chemical threats to natural ecosystems have been the tendency to let the available research tools guide the questions being addressed and a reluctance to develop entirely new tools that are designed specifically to address critical research needs. For example, our heavy reliance on single species tests is often justified by the fact that these tests have a long history of use. Arguing that we should continue to perform a test because it is a test we are good at performing is neither relevant nor sufficient justification, particularly in light of the potentially serious consequences that can result when our tests fail to provide appropriate information. Overcoming this hurdle can only be achieved 'by focusing more directly on the problems, rather than the particular intellectual tools and models used to solve them, and by ignoring arbitrary intellectual turf boundaries ... we should consider the task, evaluate existing tools' abilities to handle the job, and design new ones if the existing tools are ineffective' (Costanza *et al.*, 1991).

6.4.2 *Relevant contributions from ecology*

Recently, ecologists have been taking significant steps towards strengthening the links between theory and measurement in ecological studies (Roughgarden *et al.*, 1989; Peters, 1991). A collection of papers has been published from a meeting held in Asilomar in 1987, which brought together about 40 theoretical and empirical ecologists. *Perspectives in Ecological Theory*, published in 1989 and edited by J. Roughgarden, R.M. May and S.A. Levin, seeks to increase the dialogue between theorists and empiricists, which has been lagging in recent years to the detriment of all involved. Empiricists are realizing that they need theoretical models to simplify complex ecological systems and theorists realize that they need empiricists to tell them what they can and cannot ignore and what parameters are and are not measurable (Koehl, 1989). Ecologists have developed and tested models (incorporating both biological and physical parameters) to study the flux of nutrients among components of ecosystems (Koehl, 1989 and references therein). Such models would surely be useful to those interested in studying the fate of pollutants in natural systems. In addition, ecologists have been expending increasing efforts to develop an understanding of how individual-level processes control and are controlled by processes at the population, community and ecosystem level (Koehl, 1989). A pervasive theme throughout the book is the necessity of joining empirical and theoretical efforts and of the insights gained by linking study at various levels of biological organization. Recognition of the importance of ecosystem models to the study of artificial environmental changes is stressed throughout the book.

An excellent summary of the kind and uses of ecosystem modelling is provided in *Ecosystem Modeling in Theory and Practice: An Introduction with Case Histories*, edited by C.A.S. Hall and J.W. Day. This book, published in 1977, provides an introduction to the basic concepts and philosophies of modelling and demonstrates, with selected case histories, how models have been successfully applied to practical problems. The topics covered range from the role of value judgments in the modelling process (Friedland, 1977), modelling in the context of the law (MacBeth, 1977), specific models designed to assess environmental impact and models used for environmental management. Although published in 1977, this book is far from obsolete and should be on the required reading list for anyone involved in ecosystem modelling and risk assessment.

Two important contributions dealing with the effects of stress on natural systems have been published by Calow and Berry (1989) and Hoffmann and Parsons (1991). The former, entitled *Evolution, Ecology and Environmental Stress* is a collection of papers reprinted from the *Biological Journal of the Linnean Society*. It contains both empirical and theoretical analyses of stress and its effects at different levels of biological organization. *Evolutionary Genetics and Environmental Stress*, by Hoffmann and Parsons, provides a thorough and insightful review of the biological effects of stress, and covers molecular through quantitative genetic contributions to this area of study. The last chapter considers the implications of stress in relation to current conservation strategies and suggests a number of areas in need of study to ensure that the appropriate properties of populations or species are effectively protected.

In April 1991, the Ecological Society of America published an ecological research agenda for the 1990s. This was the first step in creating funding priorities in the face of budget tightening and a warning from the National Academy of Sciences that Congress would set priorities if scientists did not. This document emphasized the need for more basic research with relevance to pressing environmental problems such as global climate change, the loss of diversity among species and populations and the destruction of natural habitats (Gibbons, 1991). Such attention from mainstream ecologists suggests that, in the future, ecotoxicology may see greater efforts devoted toward the ecological aspects of pollutant effects.

6.4.3 Some promising theoretical approaches

Models incorporating more than one species Landis (1986) used resource–competition theory and multidimensional isocline graphical methods to examine the effects of toxicants on the dynamics of competition. The approach assumes that the toxicant of interest affects metabolic pathways used in the consumption of a resource. The models provide a useful tool for visualizing the effects of toxicant-induced shifts in resource utilization on species equilibria in simple systems with a few species. The models can also incorporate population genetic structure to explore the effects of genetic diversity on population response to toxicants, an area that has received little attention in ecotoxicological studies. This approach allows one to explore spatial and temporal heterogeneity in resources and their interactions with different toxic impacts to make predictions about competitive relationships. One very counter-intuitive result based on competition between two species was that conditions that produced the largest regions of competitive equilibria bordered situations that could lead to the extinction of one species in all areas of the resource space.

Linking individual to population response Physiological ecologists have recently developed models that express the cost of stress resistance in terms of energetic tradeoffs among growth, reproduction and survival. These models, which are solidly grounded in ecological and evolutionary theory, provide a promising approach for beginning to link individual organism performance with population dynamics in response to pollution (Koehn and Bayne, 1989; Calow and Sibly, 1990). The physiological–energetic models described in the following paragraphs are based on the balanced energy equation (Winberg, 1956). Energy balance, also called the scope for growth and reproduction (SFG), can be expressed as follows:

$$\mathrm{SFG} = P_g + P_r = A - (R_m + R_r)$$

where P_g = somatic growth; P_r = reproduction; A= absorption; R_m = maintenance metabolism; and R_r= all other metabolic costs.

Koehn and Bayne (1989) defined stress as any environmental change that acts to reduce SFG, either through a reduction in absorption (or photosynthesis in plants) or through an increase in metabolic cost. Koehn and Bayne (1989) predicted that, given a fixed energy intake (or photosynthetic rate), organisms having lower maintenance costs will have more surplus energy available for resisting environmental stress and will be able to tolerate a wider range of environmental conditions. This model predicts that, in a phenotypically variable population, those individuals exhibiting higher rates of growth and reproduction under optimal environmental conditions (due to lower maintenance metabolic costs) will be better able to tolerate stressful conditions and will retain their relative advantage, in growth and reproduction, when the population is stressed. Although this model is intuitively appealing, it ignores possible tradeoffs between growth and survival (Holloway *et al.*, 1990) and further assumes that there are no costs to stress tolerance in the absence of stress.

In contrast, Calow and Sibly (1990) noted that if the ability to resist stress carries a higher metabolic cost at all times, in other words, if the cost of stress resistance is fixed, then tolerant individuals would be at a disadvantage in the absence of stress because their higher maintenance metabolic costs would result in less energy available for growth and reproduction. Although energy balance would be expected to decrease for all individuals in the population under conditions of environmental stress, the tolerant individuals, depending on their survival increase, would now have an advantage over the susceptible individuals. A third situation, which is described by Calow and Sibly (1990) is predicted to occur when the response to stress is facultative, that is, when metabolically expensive stress-resisting processes are only deployed in the presence of a stress. In this situation, no difference in production is expected to occur between stress-tolerant and susceptible individuals in the absence of stress. However, when a stress is applied to the population, the tolerant individuals deploy resistance mechanisms, which result in an increase in metabolic cost, a decrease in energy available for growth and reproduction, but an increase in survival.

The models developed by Koehn and Bayne (1989) and Calow and Sibly (1990) lead to very general and testable hypotheses regarding the physiological and energetic responses to stress and their consequences for organism performance and population dynamics. These hypotheses can provide a conceptual framework for investigating the responses of natural populations to pollutant stress.

There is a critical need for empirical data to test the predictions of these new models and to quantify possible tradeoffs between growth and survival occurring in populations exposed to natural and anthropogenic stress. Empirical identification and quantification of the relevant tradeoffs is critical for determining to what extent physiological processes occurring at the individual organism level influence population level phenomena, such as abundance and distribution. In addition, the extent to which genetic versus environmental factors control the response of individuals and populations to pollutant stress is largely unknown. Investigations designed to incorporate population genetics and individual physiological–energetic responses to pollutant stress can provide an important step in efforts to link pollutant effects across different levels of biological organization.

Linking fate and effects Finally, there is the modelling approach we have outlined in Chapter 5 which allows rigorous and detailed examination of ecosystem-level physical, chemical and biological processes in sedimentary systems. A major advantage of this approach is that one can quantitatively model geochemical and biological processes directly at the community or ecosystem level – the level of primary interest in all ecotoxicological studies. One is then able to develop explicit and testable hypotheses related to fate/effect couplings in aquatic sedimentary ecosystems.

6.4.4 Training future ecotoxicologists

Scientific judgment has been identified as the single most important criterion in any hazard assessment decision (Dickson *et al.*, 1978). One of the most important tasks for the future is to ensure that young scientists desiring to specialize in ecotoxicology are properly trained to deal with the varied and complex problems that they are sure to face. In the current system, a typical student of ecotoxicology has either received too little training in ecotoxicological concepts or too much (at the exclusion of the necessary basics). There is little indication that the system is improving. Particularly in Europe, the trend appears to be towards increasing specialization earlier in a student's education. This is exactly the reverse of what is needed. Thus, for students who think that they might wish to pursue the study of ecotoxicology on an academic, governmental, or industrial level, we make the following suggestions.

A thorough foundation in biology and chemistry at the university level is essential. We recommend more chemistry be taken (especially physical chemistry) than is typically required for a degree in the biological sciences. More mathematics (including calculus and differential equations) should be included in the curriculum – the more the better.

A basic requirement at the graduate (i.e. MSc and PhD) level is a course in experimental design and statistics which should include multivariate statistics. A statistics course should be taken near the beginning of a student's graduate career so that he or she has adequate tools with which to design and complete an independent research project. Courses in toxicology, ecology, evolution, geochemistry and environmental management should also be required. Additional courses in the student's particular area of interest, for example, limnology, molecular biology or population genetics, should be taken as electives. Computer literacy is a must. A knowledge of economic principles would be highly beneficial for achieving insight into such ecologically relevant concepts as tradeoffs, resource limitation, gross and net gain, in addition to providing an understanding of the forces driving the public and regulatory agencies. A research project dealing with a specific problem in ecotoxicology would (and should) allow the student to integrate and put into practice the principles addressed by the formal courses. We contend that such a curriculum provides not only the necessary tools to begin to address today's problems, but that, most importantly, it provides a solid basis for facing tomorrow's as well.

6.4.5 Facilitating communication

On integrating the study of environmental problems, Cairns and Pratt (1990) commented that it is not enough that the different disciplines tolerate each other, there must be effective interaction. After listening for over 10 years to the more ecologically minded in the group condemn the irrelevance of standard toxicity tests, even the most hard-core toxicologists recognize that there are serious limitations to laboratory bioassay tests. Likewise, ecologists must concede that toxicology has been at least as successful as ecology in predicting pollutant effects.

Cairns and Pratt (1990) analysed the barriers that impede successful integration of disciplines in environmental problem solving. Concerning academic barriers, they made the following interesting observations.

> A number of tribal barriers discouraging interactions with other tribes (other disciplines) have been established. Disciplines are segregated physically on the university campus; publication in speciality journals requires both knowledge of and use of arcane and Byzantine language forms to show an awareness of the history of the speciality; and use of the latest buzz words show that a member has recently mingled with other members of the tribe.

Nowhere is this more true than in ecotoxicology. Rather than creating interdisciplinary programmes combining solid toxicological and ecological training, separate departments of ecotoxicology are popping up like mushrooms after a heavy rain (Kupchella, 1992). The supply of overspecialized journals grows daily, making it increasingly difficult for scientists in interdisciplinary fields such as ecotoxicology to keep pace with the latest advances in related areas. A scientist perusing a journal of ecotoxicology for the first time would be hard-pressed to make any sense of QSARs, MFOs, EROD, PECs, NECs, MPDs and FELS. Our poor scientist would be even more confused if he or she were to attempt to slog through a government report where attention to buzz words like 'harmonization', 'standardization' and 'state of the art' often appears to take precedence over coherent scientific discussion.

6.4.6 Some final thoughts

We must deal with the fact that many of our environmental problems are caused (or at least aggravated) by overpopulation. We worry about diminishing returns in test procedures and standardization. But will not all of our scientific efforts face diminishing returns if we fail to limit the world's human population growth? Morowitz (1991) wrote,

> No discussion of managing global habitats and preserving species can avoid the population imperative. Ignoring this problem, whether for political, ideological, or theological reasons, which is the present stance of many world political leaders, is a policy that inevitably leads to habitat loss and species depletion. There has been an unwillingness to face the conclusion that environmental planning and concern for conservation in the absence of population control are tasks worthy of the metaphor of Sisyphus.

Today, 40% of the global net primary productivity is co-opted for use by humans (Ehrlich and Wilson, 1991). The human population has been projected to double within about the next 50 years to more than 10 thousand million. Such an increase and its requisite economic growth would surely undermine even the best attempts at maintaining environmental quality. In wealthier countries, population shrinkage is either already occurring or could be achieved with little effort. Strategies to prevent the development of additional land, to improve existing facilities, to increase cleanliness while reducing waste and energy consumption, should already be priorities. Extensive and concerted efforts by the wealthier nations will also be essential for enabling poorer countries (in which many problems of global concern are likely to be concentrated) to overcome the formidable scientific, social, political and economic barriers to achieving a sustainable global environmental strategy. The really tough choices have yet to be made.

6.5 Summary and conclusions

Since the late 1970s a great deal of study has been devoted to uncovering the factors that influence chemical behaviour and toxicity in living systems. We now have a much better understanding of which intrinsic chemical properties, external environmental variables and biological factors influence toxicity, degradation and accumulation of chemicals. Yet there is much that remains to be explained, especially with regard to effects at higher levels of biological organization. The bulk of our knowledge of chemical pollutants is based on a minority of substances and species, and the call for more data is frequently heard among scientists and legislators alike. If present trends continue, there is a danger that the number of single-species laboratory tests will increase *ad nauseam*. Without a better theoretical understanding of how effects at various levels of biotic and abiotic organization interact, we will stand little chance of making sense out of even the most standardized and harmonized of data. Predictive models have been proposed, and these have been shown to predict some of the observations some of the time. There remains an urgent need for more good theoretical models of ecotoxicological phenomena. In terms of biological responses to pollutants, there are a variety of models developed by ecologists, evolutionary biologists and biogeochemists that can and should be applied. Hopefully these can then be improved by ecotoxicologists. It is clear that there are no easy answers. The problems that we face are difficult – culturally, scientifically, politically and economically. It is vitally important that ecotoxicologists be broadly trained and that management decisions, guidelines and legislation, which are presumed to be founded on scientifically sound principles, in fact are.

References

Abelson, P. H. (1992) A changing climate for scientific research. *Science*, **258,** 723.

Aldenberg, T. and Slob, W. (1991) *Confidence Limits for Hazardous Concentrations Based on Logistically Distributed NOEC Toxicity Data.* Report No. 719102002, National Institute of Public Health and Environmental Protection (RIVM), The Netherlands.

Alexander, H. C. and Quick, J. A. (1985) What industry is doing to protect aquatic life. In: *Aquatic Toxicology and Hazard Assessment: Eighth Symposium*, ASTM STP 891, (Eds R. C. Bahner and D. J. Hansen), pp. 37–44. American Society for Testing and Materials, Philadelphia.

Aller, R. C. (1980) Quantifying solute distributions in the bioturbated zone of marine sediments by defining an average microenvironment. *Geochim. Cosmochim. Acta*, **44,** 1955–65.

Aller, R. C. (1982a) Diffusion coefficients in nearshore marine sediments. *Limnol. Oceanogr.*, **27,** 552–6.

Aller, R. C. (1982b) The effects of macrobenthos on chemical properties of marine sediment and overlying water. In: *Animal–Sediment Relations: The Biogenic Alteration of Sediments*, (Eds P. L. McCall and M. J. S. Tevesz), 2nd vol. *Topics in Geobiology*, pp. 53–102. Plenum Press, New York.

Aller, R. C. (1988) Benthic fauna and biogeochemical processes in marine sediments: The role of burrow structures. In: *Nitrogen Cycling in Coastal Marine Environments*, (Eds T. H. Blackburn and J. Sφrensen), pp. 301–38. John Wiley and Sons Ltd, Chichester.

Aller, J. Y. and Aller, R. C. (1986) Evidence for localized enhancement of biological activity associated with tube and burrow structures in deep-sea sediments at the Hebble site, Western North Atlantic. *Deep Sea Res.*, **33,** 755–90.

Aller, R. C. and Cochran, J. K. (1976) ^{234}Th/^{238}U disequilibrium in the near-shore sediment: Particle reworking and diagenetic time scales. *Earth Planet. Sci. Lett.*, **29,** 37–50.

Aller, R. C. and Yingst, J. Y. (1978) Biogeochemistry of tube dwellings: A study of the sedentary polychaete *Amphitrite ornata* (Leidy). *J. Mar. Res.*, **36,** 201–54.

Alper, J. (1992) Everglades rebound from Andrew. *Science* **257,** 1852–4.

Amdur, M. O., Doull, J. and Klaassen, C. D. (Eds) (1991) *Casarett and Doull's Toxicology: The Basic Science of Poisons*, 4th edn, Pergamon Press, New York.

American Institute of Biological Sciences (1978) Criteria and rationale for decision making in aquatic hazard evaluation (third draft). In: *Estimating the Hazard of Chemical Substances to Aquatic Life*, ASTM Special Technical Publication 657, Aquatic Hazards of Pesticides Task Group of the American Institute of Biological

Sciences, (Eds J. Cairns, Jr., K. L. Dickson and A. W. Maki), pp. 241–74. American Society for Testing and Materials, Philadelphia.

Anderson, R. M. (1989) Discussion: Ecology of pests and pathogens. In: *Perspectives in Ecological Theory*, (Eds J. Roughgarden, R. M. May and S. A. Levin), pp. 348–61. Princeton University Press, Princeton.

Anderson, D. J. and Kikkawa, J. (1986) Development of concepts. In: *Community Ecology: Pattern and Process*, (Eds J. Kikkawa and D. J. Anderson), pp. 3–16. Blackwell Scientific, London.

Anholt, B. R. (1991) Measuring selection on a population of damselflies with a manipulated phenotype. *Evolution*, **45**, 1091–106.

Atkinson, H. J. (1973) The respiratory physiology of the marine nematodes *Enoplus brevis* (Bastian) and *E. communis* (Bastian). II. The effects of changes in the imposed oxygen regime. *J. Exp. Biol.*, **59**, 267–74.

Baird, D. J., Barber, I., Bradley, M. C., Soares, A. M. V. M. and Calow, P. (1989) The *Daphnia* bioassay: A critique. *Hydrobiologia*, **188/189**, 403–6.

Baird, D. J., Barber, I. and Calow, P. (1990) Clonal variation in general responses of *Daphnia magna* Straus to toxic stress. I. Chronic life-history effects. *Funct. Ecol.*, **4**, 399–408.

Baird, D. J., Barber, I., Bradley, M., Soares, A. M. V. M. and Calow, P. (1991) A comparative study of genotype sensitivity to acute toxic stress using clones of *Daphnia magna* Straus. *Ecotox. Environ. Safety*, **21**, 257–65.

Baker, J. M. and Crapp, G. B. (1974) Toxicity tests for predicting the ecological effects of oil and emulsifier pollution on littoral communities. In: *Ecological Aspects of Toxicity Testing of Oils and Dispersants*, (Eds L. R. Beynon and E. B. Cowell), pp. 23–40. Applied Science Publishers Ltd, Essex.

Bang, F. B. (1980) Monitoring pathological changes as they occur in estuaries and in the ocean in order to measure pollution (with special reference to invertebrates). *Rapp. P.-v. Réun. Cons. Int. Explor. Mer*, **179**, 118–24.

Barnthouse, L. W., O'Neill, V. O., Bartell, S. M. and Suter II, G. W. (1986) Population and ecosystem theory in ecological risk assessment. In: *Aquatic Toxicology and Environmental Fate*, 9th volume, ASTM STP 921, (Eds T. M. Poston and R. Purdy), pp. 82–96. American Society for Testing and Materials, Philadelphia.

Barrett, J. A. (1981) The evolutionary consequences of monoculture. In: *Genetic Consequences of Man Made Change*, (Eds J. A. Bishop and L. M. Cook), pp. 209–48. Academic Press, London.

Battaglia, B., Bisol, P. M., Fossato, V. U. and Rodinò, E. (1980) Studies on the genetic effects of pollution in the sea. *Rapp. P.-v. Réun. Cons. Int. Explor. Mer*, **179**, 267–74.

Bauer, J. and Capone, D. G. (1985) Effects of four aromatic organic pollutants on microbial glucose metabolism and thymidine incorporation in marine sediments. *App. Env. Microbiol.*, **49**, 828–35.

Bauer, J., Kerr, R., Bautista, M. F., Decker, C. J. and Capone, D. G. (1988) Stimulation of microbial activities and polycyclic aromatic hydrocarbon degradation in marine sediments inhabited by *Capitella capitata*. *Mar. Env. Res.*, **25**, 63–84.

Baughman, G. L. and Lassiter, R. R. (1978) Prediction of environmental pollutant concentration. In: *Estimating the Hazard of Chemical Substances to Aquatic Life*, (Eds J. Cairns, Jr., K. L. Dickson and A. W. Maki), pp. 35–54. American Society for Testing and Materials, Philadelphia.

Baughman, D. S., Moore, D. W. and Scott, G. I. (1989) A comparison and evaluation of field and laboratory toxicity tests with fenvalerate on an estuarine crustacean. *Env. Toxicol. Chem.*, **8**, 417–29.

Bayne, B. L. (1975) Aspects of physiological condition in *Mytilus edulis* L., with special reference to the effects of oxygen tension and salinity. In: *Proceedings of the Ninth European Marine Biology Symposium*, (Ed. H. Barnes), pp. 213–238. Aberdeen University Press, Aberdeen.

Bayne, B. L. (1980) Physiological measurements of stress. *Rapp. R.-v. Réun. Cons. Int. Explor. Mer*, **179,** 56–61.

Bayne, B. L., Brown, D. A., Burns, K. *et al.* (1985) *The Effects of Stress on Marine Animals*, Praeger Special Studies, New York.

Bayne, B. L. and Newell, R. C. (1983) Physiological energetics in marine molluscs. In: *The Mollusca*, vol. 4, (Eds A. S. M. Saleudden and K. M. Wilbur), pp. 407–515. Academic Press, New York.

Bell, G. (1982) *The Masterpiece of Nature: The Evolution and Genetics of Sexuality*, University of California Press, Berkeley.

Bennett, A. F. (1987) Interindividual variability: An underutilized resource. In: *New Directions in Ecological Physiology*, (Eds M. E. Feder, A. F. Bennett, W. W. Burggren and R. B. Huey), pp. 147–69. Cambridge University Press, Cambridge.

Bennett, J. (1960) A comparison of selective methods and a test of the pre-adaptation hypothesis. *Heredity*, **15,** 65–77.

Benson, W. H. and Birge, W. J. (1985) Heavy metal tolerance and metallothionein induction in fathead minnows: Results from field and laboratory investigations. *Environ. Toxicol. Chem.*, **4,** 209–17.

Berner, R. A. (1980) *Early Diagenesis: A Theoretical Approach*, Princeton University Press, Princeton.

Berry, R. J. (1980) Genes, pollution and monitoring. *Rapp. P.-v. Réun. Cons. Int. Explor. Mer*, **179,** 253–7.

Benyon, L. R. and Cowell, E. B. (Eds) (1974) *Ecological Aspects of Toxicity Testing of Oils and Dispersants*, Applied Science Publishers Ltd, Essex.

Bhat, S. G., Krishnaswami, S., Lal, D., Rama, and Moore, W. S. (1969) $^{234}Th/^{238}U$ ratios in the ocean. *Earth Planet. Sci. Lett.*, **5,** 483–91.

Bishop, J. A. and Cook, L. M. (Eds) (1981) *Genetic Consequences of Man Made Change*, Academic Press, London.

Bjerregaard, P. (1985) Effect of selenium on cadmium uptake in the shore crab *Carcinus maenas* (L.). *Aquat. Toxicol.*, **7,** 177–89.

Bjerregaard, P. (1988) Interaction between selenium and cadmium in the hemolymph of the shore crab *Carcinus maenas* (L.). *Aquat. Toxicol.*, **13,** 1–12.

Bjerregaard, P. and Vislie, T. (1986) Effect of copper on ion- and osmoregulation in the shore crab *Carcinus maenas. Mar. Biol.*, **91,** 69–76.

Blanck, H., Wallin, G. and Wängberg, S-Å. (1984) Species-dependent variation in algal sensitivity to chemical compounds. *Ecotox. Environ. Safety*, **8,** 339–351.

Blanck, H., Wängberg, S.-Å. and Molander, S. (1988) Pollution-induced community tolerance – a new ecotoxicological tool. In: *Functional Testing of Aquatic Biota for Estimating Hazards of Chemicals*, (Eds J. Cairns, Jr., and J. R. Pratt), ASTM STP 988, pp. 219–30. American Society for Testing and Materials, Philadelphia.

Bliss, C. L. (1935) The calculation of the dosage-mortality curve. *Ann. Appl. Biol.*, **22,** 134–67.

Bliss, C. L. (1957) Some principles of bioassay. *Amer. Sci.*, **45,** 449–66.

Bookhout, C. G. and Monroe, R. J. (1977) Effects of malathion on the development of crabs. In: *Physiological Responses of Marine Biota to Pollutants*, (Eds F. J. Vernberg, A. Calabrese, F. P. Thurberg and W. B. Vernberg), pp. 3–19. Academic Press, New York.

Box, G. E. P. (1976) Science and statistics. *J. Am. Stat. Soc.*, **71,** 791–9.

Bradshaw, A. D. (1984) The importance of evolutionary ideas in ecology and vice versa. In: *Evolutionary Ecology*, (Ed. B. Shorrocks), pp. 1–25. Blackwell Scientific, Oxford.

Bradshaw, A. D. and Hardwick, K. (1989) Evolution and stress – genotypic and phenotypic components. *Biol. J. Linn. Soc.*, **37,** 137–55.

Branson, D. R. (1978) Predicting the fate of chemicals in the aquatic environment from laboratory data. In: *Estimating the Hazard of Chemical Substances to Aquatic Life*, (Eds J. Cairns, Jr., K. L. Dickson and A. W. Maki), pp. 55–70. American Society for Testing and Materials, Philadelphia.

Bro-Rasmussen, F., Christiansen, K. and Folke, J. (1984) From ecotoxicology to assessment of environmental hazards of chemicals. In: *Om Miljøfarlighed og miljørisiko af Kemikalier*, (Eds J. Folke, K. Christiansen and F. Bro-Rasmussen), *Dansk Kemi* **65**(5).

Brown, A. H. D. and Burdon, J. J. (1987) Mating systems and colonizing success in plants, in *Colonization, Succession and Stability*, (Eds A. J. Gray, M. J. Crawley and P. J. Edwards), pp. 115–31. Blackwell Scientific, Oxford.

Brungs, W. A. (1986) Multispecies toxicity testing. (Book review of J. Cairns, Jr., Ed. (1985) *Multispecies Toxicity Testing*, Pergamon Press, Oxford), *BioScience*, **36,** 677–8.

Brungs, W. A. and Mount, D. I. (1978) Introduction to a discussion of the use of aquatic toxicity tests for evaluation of the effects of toxic substances. In: *Estimating the Hazard of Chemical Substances to Aquatic Life*, (Eds J. Cairns, Jr., K. L. Dickson and A. W. Maki), pp. 15–26. American Society for Testing and Materials, Philadelphia.

Bryan, G. W. (1980) Recent trends in research on heavy-metal contamination in the sea. *Helgoländer Meeresunters.*, **33,** 6–25.

Butler, G. C. (1984) Developments in ecotoxicology. *Ecol. Bull.*, **36,** 9–12.

Cairns, J., Jr. (1981) Sequential versus simultaneous testing for evaluating the hazards of chemicals to aquatic life. *Mar. Environ. Res.*, **4,** 165–6.

Cairns, J., Jr. (1984) Are single species toxicity tests alone adequate for estimating environmental hazard? *Environ. Monit. Assess.*, **4,** 259–73.

Cairns, J., Jr. (1988) What constitutes field validation of field predictions based on laboratory evidence? In: *Aquatic Toxicology and Hazard Assessment*, 10th vol., (Eds W. J. Adams, G. A. Chapman and W. G. Landis), pp. 361–68. American Society for Testing and Materials, Philadelphia.

Cairns, J., Jr. (1990) The prediction, validation, monitoring, and mitigation of anthropogenic effects upon natural systems. *Environ. Aud.*, **2,** 19–25.

Cairns, J., Jr. and Buikema, A. L., Jr. (1984) Verifying predictions of environmental safety and harm. In: *Concepts in Marine Pollution Measurements*, (Ed. H. H. White), pp. 81–112. Maryland Sea Grant College, University Maryland, College Park, MD.

Cairns, J., Jr., Dickson, K. L. and Maki, A. W. (Eds) (1978) *Estimating the Hazard of Chemical Substances to Aquatic Life*, American Society for Testing and Materials, Philadelphia.

Cairns, J., Jr. and Mount, D. I. (1990) Aquatic toxicology: Part 2 of a four-part series. *Environ. Sci. Technol.*, **24,** 154–61.

Cairns, J.Jr. and Pratt, J. R. (1986) Ecological consequence assessment: Effects of bioengineered organisms. *Water Res. Bull.*, **22,** 171–82.

Cairns, J., Jr. and Pratt, J. R. (1990) Integrating aquatic ecosystem resource management.

In: *Innovations in River Basin Management*, (Eds R. McNeil and J. E. Windsor), pp. 265–80. Canadian Water Resources Association, Cambridge, Ontario.

Cairns, J.Jr. and Pratt, J. R. (1992) (abstract) Trends in ecotoxicology. *Scientific Program of the Second European Conference on Ecotoxicology*, 11–15 May, 1992, Amsterdam, The Netherlands.

Cairns, J., Jr., Pratt, J. R., Niederlehner, B. R. and McCormick, P. V. (1986) A simple, cost-effective multispecies toxicity test using organisms with a cosmopolitan distribution. *Env. Monitor. Assess.*, **6,** 207–20.

Cairns, J., Jr. and Smith, E. P. (1989) Developing a statistical support system for environmental hazard evaluation. *Hydrobiologia*, **184,** 143–51.

Calamari, D. and Vighi, M. (1988) Quantitative structure–activity relationships in ecotoxicology: Value and limitations. In: *Proceedings of the First European Conference on Ecotoxicology*, (Eds H. Løkke, H. Tyle, and F. Bro-Rasmussen), 17–19 Oct., 1988, pp. 496–500. Conference Organizing Committee, Lyngby, Copenhagen.

Callis, C. F. (1990) Improving public understanding of science. *Environ. Sci. Technol.*, **24,** 410–11.

Calow, P. (1988) Physiological ecotoxicology: Theory, practice and application. In: *Proceedings of the First European Conference on Ecotoxicology* (Eds H. Løkke, H. Tyle, and F. Bro-Rasmussen), 17–19 Oct., 1988, pp. 23–35. Conference Organizing Committee, Lyngby, Copenhagen.

Calow, P. and Berry, R. J. (Eds) (1989) *Evolution, Ecology and Environmental Stress*, (reprinted from *Biol. J. Linnean Soc.* **37,** (1 and 2) 1989), Academic Press, London.

Calow, P. and Fletcher, C. R. (1972) A new radiotracer technique involving ^{14}C and ^{51}Cr, for estimating the assimilation efficiencies of aquatic primary consumers. *Oecologia*, **9,** 155–70.

Calow, P. and Sibly, R. M. (1990) A physiological basis of population processes: Ecotoxicological implications. *Funct. Ecol.*, **4,** 283–8.

Cammen, L. M. (1985) Metabolic loss of organic carbon by the polychaete *Capitella capitata* (Fabricius) estimated from initial weight decrease during starvation, oxygen uptake, and release of ^{14}C by uniformly labeled animals. *Mar. Ecol. Prog. Ser.*, **21,** 163–7.

Carpenter, E. J., Anderson, S. J., Harvey, G. R., Miklas, H. P. and Peck, B. B. (1972) Polystyrene spherules in coastal waters. *Science*, **178,** 749–50.

Carpenter, E. J. and Smith, K. L., Jr. (1972) Plastics on the Sargasso Sea surface. *Science*, **175,** 1240–1.

Carson, R. (1962) *Silent Spring*, Hamish Hamilton, London.

Caswell, H. (1976) The validation problem. In: *Systems Analysis and Simulation in Ecology*, Vol. 4, pp. 313–25. Academic Press, New York.

Chambers, J. M., Cleveland, W. S., Kleiner, B. and Tukey, P. A. (1983) *Graphical Methods for Data Analysis*, Statistics and Probability Series, Wadsworth and Brooks/Cole, Pacific Grove, CA.

Clark, J. R., Goodman, L. R., Borthwick, P. W. *et al.* (1989) Toxicity of pyrethroids to marine invertebrates and fish: A literature review and test results with sediment-sorbed chemicals. *Environ. Toxicol. Chem.*, **8,** 393–401.

Cleveland, W. S. (1979) Robust locally weighted regression and smoothing scatterplots. *J. Amer. Stat. Assoc.*, **74,** 829–36.

Cleveland, W. S. and McGill, R. (1984) The many faces of a scatterplot. *J. Am. Stat. Assoc.*, **79,** 807–22.

Cochran, J. K. and Aller, R. C. (1979) Particle reworking in sediments from the New

York Bight Apex: Evidence from ^{234}Th/^{238}U disequilibrium. *Est. Coast. Shelf Sci.*, **9,** 739–47.

Cochran, J. K. and Krishnaswami, S. (1980) Radium, thorium, uranium, and ^{210}Pb in deep-sea sediments and sediment pore waters from the North Equatorial Pacific. *Am. J. Sci.*, **280,** 849–99.

Cohen, J. (1988) *Statistical Power Analysis for the Behavioral Sciences*, 2nd edn, LEA Associates, Hillsdale, NJ.

Commission of the European Communities (1989) Notification of new substances in the context of Directive 67/548/EEC on the classification, packaging and labelling of new chemical substances. Guidance notes to those completing a summary notification dossier. NOTIF/15/89-final. Commission of the European Communities, Brussels.

Connell, J. H. and Sousa, W. P. (1983) On the evidence needed to judge ecological stability or persistence. *Am. Nat.*, **121,** 789–24.

Coppage, D. L. and Matthews, E. (1974) Short-term effects of organophosphate pesticides on cholinesterases of estuarine fishes and pink shrimp. *Bull. Environ. Contam. Toxicol.*, **11,** 483–8.

Costanza, R., Daly, H. E. and Bartholomew, J. A. (1991) Goals, agenda and policy recommendations for ecological economics. In: *Ecological Economics: The Science and Management of Sustainability*, (Ed R. Costanza), pp. 1–20. Columbia University Press, New York.

Cowgill, U. M. (1986) Why round-robin testing with zooplankton often fails to provide acceptable results. In: *Aquatic Toxicology and Environmental Fate*, 9th volume, ASTM STP 921, (Eds T. M. Poston and R. Purdy), pp. 349–56. American Society for Testing and Materials, Philadelphia.

Crane, M. (1990) A review of the use of aquatic multispecies systems for testing the effects of contaminants. *Report No. DoE 2587-M.* WRC plc, Henley Road, Medmenham, PO Box 16, Marlow, Buckinghamshire SL7 2HD.

Crow, J. F. (1957) Genetics of insect resistance to chemicals. *Ann. Rev. Entomol.*, **2,** 227–46.

Crow, M. E. and Taub, F. B. (1979) Designing a microcosm bioassay to detect ecosystem level effects. *Intern. J. Environ. Studies*, **13,** 141–7.

Dadd, R. H. (1975) Alkalinity within the midgut of mosquito larvae with alkaline-active digestive enzymes. *J. Insect. Physiol.*, **21,** 1847–53.

Danish Pest Infestation Laboratory (1970) *Danish Pest Inf. Lab. Ann. Rep.*, 1969, Lyngby, Denmark.

Darwin, C. (1896) *The Formation of Vegetable Mould through the Action of Worms, with Observations on their Habits.* J. Murray, London.

Davies, J. M. and Gamble, J. C. (1979) Experiments with large enclosed ecosystems. *Phil. Trans. R. Soc. Lond. B.*, **286,** 523–44.

Decho, A. W. and Luoma, S. N. (1991) Time-courses in the retention of food material in the bivalves *Potamocorbula amurensis* and *Macoma balthica*: Significance to the absorption of carbon and chromium. *Mar. Ecol. Prog. Ser.*, **78,** 303–14.

DeLaune, R. D., Patrick, W. H. and Casselman, M. E. (1981) Effect of sediment pH and redox conditions on degradation of benzo(a)pyrene. *Mar. Poll. Bull.*, **12,** 251–3.

Deneer, J. (1988) QSARs for the description and prediction of the toxicity of reactive organic chemicals to aquatic species. In: *Proceedings of the First European Conference on Ecotoxicology*, (Eds H. Løkke, H. Tyle and F. Bro-Rasmussen), 17–19 Oct., 1988, pp. 501–2. Conference Organizing Committee, Lyngby, Copenhagen.

Deneer, J. and Hermens, J. (1988) The use of quantitative structure–activity relationships in aquatic toxicology. In: *Proceedings of the First European Conference on Ecotoxicology*, (Eds H. Løkke, H. Tyle and F. Bro-Rasmussen), 17–19 Oct., 1988, pp. 447–54. Conference Organizing Committee, Lyngby, Copenhagen.

Depledge, M. H. (1992) Danish environmental research. *Mar. Poll. Bull.*, **24,** 339–42.

Depledge, M. H. and Rainbow, P. S. (1990) Models of regulation and accumulation of trace metals in marine invertebrates. *Comp. Biochem. Physiol.*, **97C,** 1–7.

Dickson, K. L., Maki, A. W., Peterson, J., Wands, R. C. and Scheier, A. (1978) Procedures for estimating hazards to aquatic life: Discussion session synopsis. In: *Estimating the Hazard of Chemical Substances to Aquatic Life*, (Eds J. Cairns, Jr., K. L. Dickson and A. W. Maki), pp. 164–87. American Society for Testing and Materials, Philadelphia.

DiToro, D. M., Mahoney, J. D., Hansen, D. J. *et al.* (1990) Toxicity of cadmium in sediments: The role of acid volatile sulfide. *Environ. Tox. Chem.*, **9,** 1487–502.

Dobzhansky, Th. (1970) *Genetics of the Evolutionary Process*, Columbia University Press, New York.

Draper, N. R. and Smith, H. (1981) *Applied Regression Analysis*, 2nd edn, John Wiley and Sons, New York.

Efron, B. and Tibshirani, R. (1991) Statistical data analysis in the computer age. *Science*, **253,** 390–5.

Ehrlich, P. R. and Wilson, E. O. (1991) Biodiversity studies: Science and policy. *Science*, **253,** 758–62.

Elliot, M. (1991) Coastal Management (Book Review). *Mar. Pollut. Bull.*, **22,** 213–14.

Emery, R. M. and Mattson, G. G. (1986) Systems ecology and environmental law: Do they speak the same language? In: *Aquatic Toxicology and Environmental Fate*, 9th volume, ASTM STP 921, (Eds T. M. Poston and R. Purdy), pp. 25–41. American Society for Testing and Materials, Philadelphia.

Endler, J. A. (1986) *Natural Selection in the Wild.* Princeton University Press, Princeton, NJ.

Ernst, W. (1980) Effects of pesticides and related organic compounds in the sea. *Helgoländer Meeresunters.*, **33,** 301–12.

Erwin, T. L. (1991) An evolutionary basis for conservation strategies. *Science*, **253,** 750–2.

Falconer, D. S. (1981) *Introduction to Quantitative Genetics*, 2nd edn, Longman, London.

Famme, P. and Kofoed, L. H. (1982) Rates of carbon release and oxygen uptake by the mussel, *Mytilus edulis* L., in response to starvation and oxygen. *Mar. Biol. Lett.*, **3,** 241–56.

Farrington, J. W. (1980) An overview of the biogeochemistry of fossil fuel hydrocarbons in the marine environment. In: *Petroleum in the Marine Environment*, (Eds L. Petrakis and F. T. Weiss), *Advances in Chemistry Series*, no. 185, ACS, Washington, DC.

Feder, M. E., Bennett, A. F., Burggren, W. W. and Huey, R. B. (Eds) (1987) *New Directions in Ecological Physiology*, Cambridge University Press, Cambridge.

Finney, D. J. (1971) *Probit Analysis*, 3rd edn, Cambridge University Press, Cambridge.

Finney, D. J. (1978) *Statistical Method in Biological Assay*, 3rd edn, Charles Griffen and Co, London.

Fisher, N. S. (1977) On the differential sensitivity of estuarine and open-ocean diatoms to exotic chemical stress. *Am. Nat.*, **111,** 871–95.

Fisher, N. S., Graham, L. B., Carpenter, E. J. and Wurster, C. F. (1973) Geographic differences in phytoplankton sensitivity to PCBs. *Nature*, **241,** 548–9.

Fisher, R. A. (1932) *Statistical Methods for Research Workers*. 4th edn, Oliver and Boyd, Edinburgh.

Forbes, T. L. (1989) The importance of size-dependent physiological processes in the ecology of the deposit-feeding polychaete *Capitella* species 1. *PhD Dissertation*, State University of New York, Stony Brook, NY.

Forbes, T. L. (1992) The design and analysis of concentration–response experiments. In: *Handbook of Ecotoxicology*, (Ed P. Calow), pp. 438–60. Blackwell Scientific, Oxford.

Forbes, T. L. and Lopez, G. R. (1990) The effect of food concentration, body size, and environmental oxygen tension on the growth of the deposit-feeding polychaete, *Capitella* species 1. *Limnol. Oceanogr.*, **35,** 1535–44.

Forbes, T. L. and Forbes, V. E. (1993) A critique of the use of distribution-based extrapolation models in ecotoxicology. *Funct. Ecol.* **7**, 249–54.

Forbes, V. E. and Depledge, M. H. (1992a) Population response to pollutants: The significance of sex. *Funct. Ecol.* **6,** 376–81.

Forbes, V. E. and Depledge, M. H. (1992b) Cadmium effects on the carbon and energy balance of mudsnails. *Mar. Biol.* **113,** 263–9.

Forbes, V. E. and Depledge, M. H. (1993) Testing versus research in ecotoxicology: a response to Baird and Calow. *Funct. Ecol.* **7** (in press).

Forbes, V. E. and Lopez, G. R. (1989) The role of sediment particle size in the nutritional energetics of a surface deposit-feeder. II. Energetic cost measured as ^{14}C loss from uniformly labeled *Hydrobia truncata* (Vanatta). *J. Exp. Mar. Biol. Ecol.*, **126,** 193–202.

Forbes, V. E. and Lopez, G. R. (1990) The role of sediment type in growth and fecundity of mud snails (Hydrobiidae). *Oecologia*, **83,** 53–61.

Fossi, M. C., Leonzio, C., Focardi, S., Lari, L. and Renzoni, A. (1991) Modulation of mixed-function oxidase activity in Black-headed Gulls living in anthropic environments: Biochemical acclimatization or adaptation? *Environ. Toxicol. Chem.*, **10,** 1179–88.

Fossi, C., Leonzio, C., Focardi, S. and Renzoni, A. (1988) The Black-headed Gull's adaptation to polluted environments: The role of the mixed-function oxidase detoxication system. *Environ. Conserv.*, **15,** 221–4.

Frederick, R. J. and Pilsucki, R. W. (1991) Nontarget species testing of microbial products intended for use in the environment. In: *Risk Assessment in Genetic Engineering*, (Eds M. A. Levin and H. S. Strauss), pp. 32–50. McGraw-Hill, New York.

Freij, L. (Ed.) (1991) Seminar on Environmental Classification and Labelling of Chemicals, EC-EFTA Meeting, Uppsala, Sweden, March 20–21, Printgraf, Stockholm.

Friedland, E. I. (1977) Values and environmental modeling. In: *Ecosystem Modeling in Theory and Practice: An Introduction with Case Histories*, (Eds C. A. S. Hall and J. W. Day, Jr.), pp. 115–31. John Wiley and Sons, New York.

Futoma, D. J., Smith, S. R., Smith, T. E., and Tanaka, J. (1981) *Polycyclic Aromatic Hydrocarbons in Water Systems*, CRC Press, Boca Raton, FL.

Futuyma, D. J. (1979) *Evolutionary Biology*, Sinauer Associates, Sunderland.

Gaddum, J. H. (1933) Reports on biological standards. III. Methods of biological assay depending on a quantal response. *Med. Res. Counc., Spec. Rep. Ser.*, no. 183.

Gallo, M. A. and Doull, J. (1991) History and scope of toxicology. In: *Casarett and Doull's Toxicology: The Basic Science of Poisons*, 4th edn, (Eds M. O. Amdur, J. Doull and C. D. Klaassen), pp. 3–11. Pergamon Press, New York.

Gardner, W. S., Lee, R. F., Tenore, K. R. and Smith, L. W. (1979) Degradation of selected polycyclic aromatic hydrocarbons in coastal sediments: Importance of microbes and polychaete worms. *Water Air Soil Pollut.*, **11,** 339–47.

Gibbons, A. (1991) Ecologists set broad priorities for 1990s. *Science*, **252,** 504.

Giddings, J. M. (1986) Protecting aquatic resources: An ecologist's perspective. In: *Aquatic Toxicology and Environmental Fate*, 9th volume, ASTM STP 921, (Eds T. M. Poston and R. Purdy), pp. 97–106. American Society for Testing and Materials, Philadelphia.

Giddings, J. M. and Eddlemon, G. K. (1979) Some ecological and experimental properties of complex aquatic microcosms. *Int. J. Environ. Stud.*, **13,** 119–23.

Gilfillan, E. S. (1980) The use of scope-for-growth measurements in monitoring petroleum pollution. *Rapp. P. -v. Réun. Cons. Int. Explor. Mer*, **179,** 71–5.

Gnaiger, E. (1983) Heat dissipation and energetic efficiency in animal anoxibiosis: Economy contra power. *J. Exp. Zool.*, **228,** 471–90.

Goyer, R. A. (1991) Toxic effects of metals. In: *Casarett and Doull's Toxicology: The Basic Science of Poisons*, (Eds M. O. Amdur, J. Doull, and C. D. Klaassen), pp. 623–80. Pergamon Press, New York.

Granby, K. (1987) Levels of hydrocarbons and chlorinated compounds in the Danish Sea areas, 1985–1986. *Rep. Mar. Pollut. Lab.*, No. 12, Marine Pollution Laboratory, Copenhagen, Denmark.

Grant, J. and Cranford, P. J. (1991) Carbon and nitrogen scope for growth as a function of diet in the sea scallop *Placopecten magellanicus. J. Mar. Biol. Assoc. UK*, **71,** 437–50.

Grassle, J. F. and Grassle, J. P. (1974) Opportunistic life histories and genetic systems in marine benthic polychaetes. *J. Mar. Res.*, **32,** 253–84.

Grassle, J. P. and Grassle, J. F. (1976) Sibling species of the marine pollution indicator *Capitella* (Polychaeta). *Science*, **192,** 567–9.

Gray, J. S. (1979) Pollution-induced changes in populations. *Phil. Trans. Roy. Soc. Lond. B*, **286,** 545–61.

Gray, J. S. (1980) The measurement of effects of pollutants on benthic communities. *Rapp. P.-v. Réun. Cons. Int. Explor. Mer*, **179,** 188–93.

Gray, J. S. (1989) Effects of environmental stress on species rich assemblages. *Biol. J. Linn. Soc.*, **37,** 19–32.

Gray, J. S. (1991) Climate change. *Mar. Pollut. Bull.*, **22,** 169–71.

Gray, J. S., Aschan, M., Carr, M. R. *et al.* (1988) Analysis of community attributes of the benthic macrofauna of Frierfjord/Langesundfjord and in a mesocosm experiment. *Mar. Ecol. Prog. Ser.*, **46,** 151–65.

Gray, J. S., Clarke, K. R., Warwick, R. M. and Hobbs, G. (1990) Detection of initial effects of pollution on marine benthos: An example from the Ekofisk and Eldfisk oilfields, North Sea. *Mar. Ecol. Prog. Ser.*, **66,** 285–99.

Gray, J. S. and Pearson, T. H. (1982) Objective selection of sensitive species indicative of pollution-induced change in benthic communities. I. Comparative methodology. *Mar. Ecol. Prog. Ser.*, **9,** 111–19.

Green, R. H. (1979) *Sampling Design and Statistical Methods for Environmental Biologists.* John Wiley and Sons, New York.

Gross, L. J. (1989) Plant physiological ecology: A theoretician's perspective. In: *Perspectives in Ecological Theory*, (Eds J. Roughgarden, R. M. May and S. A. Levin), pp. 11–24. Princeton University Press, Princeton.

Gschwend, P. M. and Hites, R. A. (1981) Fluxes of polycyclic aromatic hydrocarbons to marine and lacustrine sediments in the northeastern United States. *Geochim. Cosmochim. Acta*, **45,** 2359–67.

Guinasso, N. L., Jr. and Schink, D. R. (1975) Quantitative estimates of biological mixing rates in abyssal sediments. *J. Geophys. Res.*, **80,** 3032–43.

Hahne, H. C. H. and Kroontje, W. (1973) Significance of pH and chloride concentration on behavior of heavy metal pollutants: Mercury (II), cadmium (II), zinc (II), and lead (II). *J. Environ. Qual.*, **2,** 444–50.

Hambrick, G. A., III, DeLaune, R. D. and Patrick, W. H., Jr. (1980) Effect of estuarine sediment pH and oxidation–reduction potential on microbial hydrocarbon degradation. *Appl. Env. Microbiol.*, **40,** 365–9.

Harrison, R. G. (1977) Parallel variation at an enzyme locus in sibling species of field crickets. *Nature*, **266,** 168–170.

Hart, J. W. and Jensen, N. J. (1992) Integrated risk assessment or integrated risk management? *Regul. Toxicol. Pharmacol.*, **15,** 32–40.

Hartley, J. P. (1984) The benthic ecology of the Forties Oilfield (North Sea). *J. Exp. Mar. Biol. Ecol.*, **80,** 161–95.

Hayes, W. J., Jr. (1991) Introduction. In: *Handbook of Pesticide Toxicology*, Vol. 1, (Eds W. J. Hayes, Jr. and E. R. Laws, Jr.), pp. 1–38. Academic Press, San Diego, CA.

Hayes, W. J., Jr. and Laws, E. R. (1991) *Handbook of Pesticide Toxicology*, Academic Press, San Diego, CA.

Hilbish, T. J., Deaton, L. E. and Koehn, R. K. (1982) Effect of an allozyme polymorphism on regulation of cell volume. *Nature*, **298,** 688–9.

Hilbish, T. J. and Koehn, R. K. (1985a) Dominance in physiological and fitness phenotypes at an enzyme locus. *Science*, **229,** 52–4.

Hilbish, T. J. and Koehn, R. K. (1985b) The physiological basis of natural selection at the *Lap* locus. *Evolution*, **39,** 1302–47.

Hobbs, G. (1987) Environmental survey of the benthic sediments of the Ekofisk and Eldfisk oilfields, July 1987. *Oil Pollution Research Unit Report No. FSC/OPRU/24/87*, Oil Pollution Research Unit, Field Studies Council, Orielton Field Centre, Pembroke, Dyfed SA71 5EZ, UK.

Hoffmann, A. A. and Parsons, P. A. (1991) *Evolutionary Genetics and Environmental Stress*, Oxford Science Publications, Oxford.

Hollander, M., Wolfe, D. A. (1973) *Nonparametric Statistical Methods*, John Wiley & Sons, New York.

Holloway, G. J., Sibly, R. M. and Povey, S. R. (1990) Evolution in toxin-stressed environments. *Funct. Ecol.*, **4,** 289–94.

Hoskins, W. M. and Gordon, H. T. (1956) Arthropod resistance to chemicals. *Ann. Rev. Ent.*, **1,** 89–122.

Houghton, R. A. (1990) The global effects of tropical deforestation. *Environ. Sci. Technol.*, **24,** 414–22.

Hutcheson, M., Miller, D. C. and White, A. Q. (1985) Respiratory and behavioral responses of the grass shrimp *Palaemonetes pugio* to cadmium and reduced dissolved oxygen. *Mar. Biol.*, **88,** 59–66.

Hutchings, P. (1991) Coastal zone issues in Australia. *Mar. Pollut. Bull.*, **22,** 220–1.

Hylleberg, J. (1975) Selective feeding by *Abarenicola pacifica* with notes on *Abarenicola vagabunda* and a concept of gardening in lugworms. *Ophelia*, **14,** 113–37.

Ivanovici, A. M. (1980a) Adenylate energy charge: An evaluation of applicability to assessment of pollution effects and directions for future research. *Rapp. P.-v. Réun. Cons. Int. Explor. Mer*, **179,** 23–8.

Ivanovici, A. M. (1980b) Application of adenylate energy charge to problems of environmental impact assessment in aquatic organisms. *Helgoländer Meeresunters.*, **33,** 556–65.

Jensen, K. and Jørgensen, K. F. (1984) Sources and effects of petroleum hydrocarbon pollution in the Baltic Sea. *Ophelia*, Suppl. 3, 61–8.

Johnson, R. G. (1974) Particulate matter at the sediment-water interface in coastal environments. *J. Mar. Res.*, **32,** 313–30.

Jop, K. M., Rodgers, J. H., Jr., Dorn, P. B. and Dickson, K. L. (1986) Use of hexavalent chromium as a reference toxicant in aquatic toxicity tests. In: *Aquatic Toxicology and Environmental Fate*, 9th vol., ASTM STP 921, (Eds T. M. Poston and R. Purdy), pp. 390–403. American Society for Testing and Materials, Philadelphia.

Kareiva, P. (1989) Renewing the dialogue between theory and experiments in population ecology. In: *Perspectives in Ecological Theory*, (Eds J. Roughgarden, R. M. May and S. A. Levin), pp. 68–88. Princeton University Press, Princeton.

Kaufmann, K. W. (1981) Fitting and using growth curves. *Oecologia* (Berl.), **49,** 293–9.

Kemp, W. M., Smith, W. H. B., McKellar, H. N. *et al.* (1977) Energy cost-benefit analysis applied to power plants near Crystal River, Florida. In: *Ecosystem Modeling in Theory and Practice: An Introduction With Case Histories*, (Eds C. A. S. Hall and J. W. Day, Jr.), pp. 507–44. John Wiley and Sons, New York.

Kenaga, E. E. and Lamb, D. W. (1981) Introduction. In: *Avian and Mammalian Wildlife Toxicology: Second Conference*, ASTM Special Technical Publication 757, (Eds D. W. Lamb and E. E. Kenaga), pp. 1–3. American Society for Testing and Materials, Philadelphia.

Kersting, K. (1984) Development and use of an aquatic micro-ecosystem as a test system for toxic substances. Properties of an aquatic micro-ecosystem IV. *Int. Rev. Ges. Hydrobiol.*, **69,** 567–607.

Kersting, K. (1985) Properties of an aquatic micro-ecosystem V. Ten years of observations of the prototype. *Verh. Internat. Verein. Limnol.*, **22,** 3040–5.

Kikkawa, H. (1964) Genetic studies on the resistance to parathion in *Drosophila melanogaster*. II. Induction of a resistance gene from its susceptible allele. *Botyu-kagaku*, **29,** 37–42.

Kimerle, R. A., Gledhill, W. E. and Levinskas, G. J. (1978) Environmental safety assessment of new materials. In: *Estimating the Hazard of Chemical Substances to Aquatic Life*, ASTM Special Technical Publication 657, (Eds J. Cairns, Jr., K. L. Dickson and A. W. Maki), pp. 132–46. American Society for Testing and Materials, Philadelphia.

Kingston, P. F. (1987) Field effects of platform discharges on benthic macrofauna. *Phil. Trans. R. Soc. Lond. B*, **316,** 545–65.

Kinne, O. (1968) International symposium 'Biological and hydrographical problems of water pollution in the North Sea and adjacent waters': Closing address. *Helgoländer Wiss. Meeresunters.*, **17,** 518–22.

Kinne, O. (1980) 14th European Marine Biology Symposium 'Protection of life in the sea': Summary of symposium papers and conclusions. *Helgoländer Meeresunters.*, **33,** 732–61.

Kittredge, J. S. (1980) Behavioral bioassays: The range of response thresholds. *Rapp. P.-v. Réun. Cons. Int. Explor. Mer*, **179,** 152–3.

Klaassen, C. D. and Eaton, D. L. (1991) Principles of toxicology. In: *Casarett and Doull's Toxicology: The Basic Science of Poisons*, 4th edn, (Eds M. O. Amdur, J. Doull and C. D. Klaassen), pp. 12–49. Pergamon Press, New York.

Klaassen, C. D. and Rozman, K. (1991) Absorption, distribution, and excretion of toxicants. In: *Casarett and Doull's Toxicology: The Basic Science of Poisons*, 4th edn, (Eds M. O. Amdur, J. Doull and C. D. Klaassen), pp. 50–87. Pergamon Press, New York.

Klerks, P. L. (1987) Adaptation to metals in benthic macrofauna. PhD Dissertation, State University of New York, Stony Brook, NY.

Klerks, P. L. and Levinton, J. S. (1989) Effects of heavy metals in a polluted aquatic ecosystem. In: *Ecotoxicology: Problems and Approaches*, (Eds S. A. Levin, M. A. Harwell, J. R. Kelly and K. D. Kimball), pp. 41–67. Springer-Verlag, Berlin.

Koehl, M. A. R. (1989) Discussion: From individuals to populations. In: *Perspectives in Ecological Theory*, (Eds J. Roughgarden, R. M. May and S. A. Levin), pp. 39–53. Princeton University Press, Princeton.

Koehn, R. K. and Bayne, B. L. (1989) Towards a physiological and genetical understanding of the energetics of the stress response. *Biol. J. Linn. Soc.*, **37**, 157–71.

Koehn, R. K., Newell, R. I. E. and Immerman, F. (1980) Maintenance of an amino peptidase allele frequency cline by natural selection. *Proc. Natl. Acad. Sci. USA*, **77**, 5385–9.

Koehn, R. K., Zera, A. J. and Hall, J. G. (1983) Enzyme polymorphism and natural selection. In: *Evolution of Genes and Proteins*, (Eds M. Nei and R. K. Koehn), pp. 115–36. Sinauer, Sunderland.

Kofoed, L. H. (1975) The feeding biology of *Hydrobia ventrosa* (Montagu). II. Allocation of the components of the carbon-budget and the significance of the secretion of dissolved organic material. *J. Exp. Mar. Biol. Ecol.*, **19**, 243–56.

Kooijman, S. A. L. M. (1987) A safety factor for LC_{50} values allowing for differences in sensitivity among species. *Wat. Res.*, **21**, 269–76.

Koshland, D. E., Jr. (1991) Frontiers in biotechnology. *Science*, **252**, 1593.

Kristensen, E., Jensen, M. H. and Andersen, T. K. (1985) The impact of polychaete (*Nereis virens* Sars) burrows on nitrification and nitrate reduction in estuarine sediments. *J. Exp. Mar. Biol. Ecol.*, **85**, 75–91.

Kuiper-Goodman, T. (1989) Risk assessment of mycotoxins. In: *Mycotoxins and Phycotoxins '88*, A collection of invited papers presented at the Seventh International IUPAC Symposium on Mycotoxins and Phycotoxins, 6–19 August, Tokyo, Japan, 1988, (Eds S. Natori, K. Hashimoto and Y. Ueno), pp. 257–64. Elsevier, Amsterdam.

Kupchella, C. E. (1992) Education of environmental specialists and generalists in American universities. In: *The Science of Global Change*, (Eds D. A. Dunnette and R. J. O'Brien), pp. 473–80. American Chemical Society.

Lampert, W. (1977) Studies on the carbon balance of *Daphnia pulex* as related to environmental conditions. I. Methodological problems of the use of ^{14}C for the measurement of carbon assimilation. *Arch. Hydrobiol. Suppl.*, **48**, 287–309.

Lampert, W., Fleckner, W., Pott, E., Schober, U. and Störkel, K-U. (1989) Herbicide effects on planktonic systems of different complexity. *Hydrobiologia*, **188/189**, 415–24.

Landis, W. G. (1986) Resource competition modelling of the impacts of xenobiotics on biological communities. In: *Aquatic Toxicology and Environmental Fate*, 9th vol., ASTM STP 921, (Eds T. M. Poston and R. Purdy), pp. 55–72. American Society for Testing and Materials, Philadelphia.

Lee, C. M. (1985) Scientific considerations in the international arena. In: *Aquatic Toxicology and Hazard Assessment*, 8th Symposium, ASTM STP 891, (Eds R. C. Bahner and D. J. Hansen), pp. 15–26. American Society for Testing and Materials, Philadelphia.

Lee, C. M., Turner, C. A. and Huntington, E. (1986) Factors affecting the culture of *Daphnia magna*. In: *Aquatic Toxicology and Environmental Fate*, 9th vol. ASTM STP 921, (Eds T. M. Poston and R. Purdy), pp. 357–68. American Society for Testing and Materials, Philadelphia.

Lee, R., Davies, J. M., Freeman, H. C., Ivanovici, A., Moore, M. N., Stegeman, J. and Uthe, J. F. (1980) Biochemical techniques for monitoring biological effects of pollution in the sea. *Rapp. P. -v. Réun. Cons. Int. Explor. Mer*, **179,** 48–55.

Lee, R. F., Takahaski, M., Beers, J. R. *et al.* (1977) Controlled ecosystems: Their use in the study of the effects of petroleum hydrocarbons on plankton. In: *Physiological Responses of Marine Biota to Pollutants*, (Eds F. J. Vernberg, A. Calabrese, F. P. Thurberg and W. B. Vernberg), pp. 323–342. Academic Press, New York.

Leland, H. V. and Kuwabara, J. S. (1985) Trace Metals. In: *Fundamentals of Aquatic Toxicology*, (Eds G. M. Rand and S. R. Petrocelli), pp. 374–415. Hemisphere, New York.

Lewis, J. R. (1980) Options and problems in environmental management and evaluation. *Helgoländer Meeresunters.*, **33,** 452–66.

Lewontin, R. C. (1965) Selection for colonizing ability. In: *The Genetics of Colonizing Species*, (Eds H. G. Baker and G. L. Stebbins), pp. 77–94. Academic Press, New York.

Lindzen, R. S. (1990) Some remarks on global warming. *Environ. Sci. Technol.*, **24,** 424–5.

Lipnick, R. L. (1985) Research needs in developing structure activity relationships. In: *Aquatic Toxicology and Hazard Assessment*, 8th Symposium, ASTM STP, 891, (Eds R. C. Bahner and D. J. Hansen), pp. 78–82. American Society for Testing and Materials, Philadelphia.

Lloyd, R. (1980) Toxicity testing with aquatic organisms: A framework for hazard assessment and pollution control. *Rapp. P. -v. Réun. Cons. Int. Explor. Mer*, **179,** 339–41.

Loehle, C. (1987) Hypothesis testing in ecology: Psychological aspects and the importance of theory maturation. *Quart. Rev. Biol.*, **62,** 397–409.

Loomis, T. A. (1975) Acute and prolonged toxicity tests. *J. Assoc. Official Analytical Chem.*, **58,** 645.

Lopez, G. R. and Crenshaw, M. A. (1982) Radiolabelling of sedimentary organic matter with ^{14}C-formaldehyde: Preliminary evaluation of a new technique for use in deposit-feeding studies. *Mar. Ecol. Prog. Ser.*, **8,** 283–9.

Lopez, G. R. and Levinton, J. S. (1987) Ecology of deposit-feeding animals in marine sediments. *Quart. Rev. Biol.*, **62,** 235–60.

Lu, J. C. S. and Chen, K. Y. (1977) Migration of trace metals in interfaces of seawater and polluted surficial sediments. *Env. Sci. Technol.*, **11,** 174–82.

Luoma, S. N. (1977) Detection of trace contaminant effects in aquatic ecosystems. *J. Fish. Res. Board Can.*, **34,** 436–9.

Luria, S. and Delbrück, M. (1943) Mutations of bacteria from virus sensitivity to virus resistance. *Genetics*, **28,** 491–511.

Lynch, M., Weider, L. J. and Lampert, W. (1986) Measurement of carbon balance in *Daphnia. Limnol. Oceanogr.*, **31,** 17–33.

Macnair, M. B. (1981) Tolerances of higher plants to toxic materials. In: *Genetic Consequences of Man made Change* (Eds J. A. Bishop and L. M. Cook). Academic Press, London.

MacBeth, A. (1977) Modeling in the Context of the Law. In: *Ecosystem Modeling in Theory and Practice: An Introduction with Case Histories*, (Eds C. A. S. Hall and J. W. Day, Jr.), pp. 197–210. John Wiley and Sons, New York.

Maki, A. W. and Bishop, W. E. (1985) Chemical safety evaluation. In: *Fundamentals of Aquatic Toxicology*, (Eds G. M. Rand and S. R. Petrocelli), pp. 619–35. Hemisphere, New York.

Maki, A. W., Dickson, K. L. and Cairns, J., Jr. (1979) Introduction. In: *Analyzing the Hazard Evaluation Process*, (Eds K. L. Dickson, A. W. Maki and J. Cairns, Jr.), pp. 1–6. American Fisheries Society, Bethesda, MD.

Maki, A. W. and Duthie, J. R. (1978) Summary of proposed procedures for the evaluation of aquatic hazard. In: *Estimating the Hazard of Chemical Substances to Aquatic Life*, (Eds J. Cairns, Jr., K. L. Dickson, and A. W. Maki), pp. 153–63. American Society for Testing and Materials, Philadelphia.

Mann, C. C. (1991) Extinction: Are ecologists crying wolf? *Science*, **253,** 736–8.

Matisoff, G. (1982) Mathematical models of bioturbation. In: *Animal–Sediment Relations: The Biogenic Alteration of Sediments*, (Eds P. L. McCall and M. J. S. Tevesz), 2nd volume, *Topics in Geobiology*, pp. 289–330. Plenum Press, New York.

Maurer, B. A. (1987) Scaling of biological community structure: A systems approach to community complexity. *J. Theor. Biol.*, **127,** 97–110.

May, R. (1973) *Stability and Complexity in Model Ecosystems*, Princeton University Press, Princeton.

Maynard Smith, J. (1989) *Evolutionary Genetics*, Oxford University Press, Oxford.

McCarthy, J. F. and Shugart, L. R. (1990) *Biomarkers of Environmental Contamination*, Lewis Publishers, Boca Raton, FL.

McKenney, C. L. and Hamaker, D. B. (1984) Effects of fenvalerate on larval development of *Palaemonetes pugio* (Holthuis) and on larval metabolism during osmotic stress. *Aquat. Toxicol.*, **5,** 343–55.

McKenney, C. L. and Matthews, E. (1990) Alterations in the energy metabolism of an estuarine mysid (*Mysidopsis bahia*) as indicators of stress from chronic pesticide exposure. *Mar. Environ. Res.*, **30,** 1–19.

McManus, J. W. and Pauly, D. (1990) Measuring ecological stress: Variations on a theme by R. M. Warwick. *Mar. Biol.*, **106,** 305–8.

McNeilly, T. (1968) Evolution in closely adjacent plant populations. III. *Agrostis tenuis* on a small copper mine. *Heredity*, **23,** 99–108.

McNeilly, T. and Bradshaw, A. D. (1968) Evolutionary processes in populations of copper tolerant *Agrostis tenuis* Sibth. *Evolution*, **22,** 108–18.

Meadows, P. S. and Tufail, A. (1986) Bioturbation, microbial activity and sediment properties in an estuarine ecosystem. *Proc. Roy. Soc. Edinburgh*, **90B,** 129–42.

Medawar, P. B. (1979) *Advice to a Young Scientist*. Harper and Row, New York.

Menzel, D. W. (1977) Summary of experimental results: Controlled ecosystem pollution experiment. *Bull. Mar. Sci.*, **27,** 142–5.

Menzer, R. E. (1991) Water and soil pollutants. In: *Casarett and Doull's Toxicology: The Basic Science of Poisons*, (Eds M. O. Amdur, J. Doull and C. D. Klaassen), pp. 872–902. Pergamon Press, New York.

Milbrath, L. W. (1991) Why we must turn away from environmental chemicals. *Envir. Carcino. Revs. (J. Envir. Sci. Hlth.)*, **C8(2),** 267–75.

Moore, M. N. (1980) Cytochemical determination of cellular responses to environmental stressors in marine organisms. *Rapp. P. -v. Réun. Cons. Int. Explor. Mer*, **179,** 7–15.

Moriarty, F. (1983) *Ecotoxicology: The Study of Pollutants in Ecosystems*, Academic Press, London.

Moriarty, F. (1984) Persistent contaminants, compartmental models and concentration along food-chains. *Ecol. Bull.*, **36,** 35–45.

Morowitz, H. J. (1991) Balancing species preservation and economic considerations. *Science*, **253,** 752–4.

Mount, D. I. (1979) Adequacy of laboratory data for protecting aquatic communities,

In: *Analyzing the Hazard Evaluation Process*, (Eds K. L. Dickson, A. W. Maki and J. Cairns, Jr.), pp. 112–18. American Fisheries Society, Bethesda, MD.

Mueller, L. D., Guo, P. and Ayala, F. J. (1991) Density-dependent natural selection and trade-offs in life history traits. *Science*, **253**, 433–5.

National Oceanic and Atmospheric Administration (1989) An evaluation of candidate measures of biological effects for the National Status and Trends Program. *NOAA Technical Memorandum NOS OMA 45*, National Oceanic and Atmospheric Administration, Seattle, WA.

National Oceanic and Atmospheric Administration (1991) *National Ocean Pollution Plan, Federal Plan for Ocean Pollution Research, Development, and Monitoring, Fiscal Years 1992–1996*, US Department of Commerce, National Oceanic and Atmospheric Administration, Washington, DC.

Neff, J. M. (1979) *Polycyclic Aromatic Hydrocarbons in the Aquatic Environment: Sources, Fates and Biological Effects*. Applied Science, London.

Neff, J. M. (1985) Polycyclic aromatic hydrocarbons. In: *Fundamentals of Aquatic Toxicology*, (Eds G. M. Rand and S. R. Petrocelli), pp. 416–54. Hemisphere, New York.

Neff, J. M. and Anderson, J. W. (1981) *Responses of Marine Animals to Petroleum and Specific Petroleum Hydrocarbons*. Applied Science, London.

Nendza, M. and Klein, W. (1990) Comparative QSAR study on freshwater and estuarine toxicity. *Aquat. Toxicol.*, **17**, 63–74.

Neuhold, J. M. (1986) Toward a meaningful interaction between ecology and aquatic toxicology. In: *Aquatic Toxicology and Environmental Fate*, 9th volume, ASTM STP 921, (Eds T. M. Poston and R. Purdy), pp. 11–24. American Society for Testing and Materials, Philadelphia.

Nevo, E., Lavie, B. and Ben-Shlomo, R. (1983) Selection of allelic isozyme polymorphisms in marine organisms: Pattern, theory, and application. In: *Isozymes: Current Topics in Biological and Medical Research*, Vol. 10, pp. 69–92. Alan R. Liss: New York.

Niederlehner, B. R., Pratt, J. R., Buikema, A. L., Jr. and Cairns, J., Jr. (1986) Comparison of estimates of hazard derived at three levels of complexity. In: *Community Toxicity Testing*, ASTM STP 920, (Ed J. Cairns, Jr.), pp. 30–48. American Society for Testing and Materials, Philadelphia.

Nimmo, D. R. (1985) Pesticides. In: *Fundamentals of Aquatic Toxicology*, (Eds G. M. Rand and S. R. Petrocelli), pp. 335–73. Hemisphere, New York.

Nordic Council of Ministers (1990) Substances and preparations dangerous for the environment: A system for classification, labelling and safety data sheets. Final report from a Nordic Working Group. *Environmental Report* 1990:10E, Nordic Council of Ministers, Copenhagen, Denmark.

North Sea Conference (1990) Interim report on the quality status of the North Sea. Report from Third International Conference on the Protection of the North Sea. Ministry of Transport and Public Works, The Hague, The Netherlands.

Olla, B. L., Pearson, W. H. and Studholme, A. L. (1980) Applicability of behavioral measures in environmental stress assessment. *Rapp. P. -v. Réun. cons. Int. Explor. Mer*, **179**, 162–73.

Organization for Economic Cooperation and Development (1981) *Guidelines for the Testing of Chemicals*, Organization for Economic Cooperation and Development, Paris, France (including all updates to 1991).

Organization for Economic Cooperation and Development (1989) Report of the OECD workshop on ecological effects assessment. *OECD Environmental Monographs*,

No. 26, Organization for Economic Cooperation and Development, Paris, France.

Organization for Economic Cooperation and Development (1991) Draft report of the OECD workshop on the extrapolation of laboratory aquatic toxicity data to the real environment. Organization for Economic Cooperation and Development, Paris, France.

O'Neill, R. V., De Angelis, D. L., Waide, J. B. and Allen, T. F. H. (1986) *A Hierarchical Concept of Ecosystems*, Princeton University Press, Princeton.

Ornstein, R., Ehrlich, P. (1989) *New World, New Mind: Changing the Way we Think to Save our Future.* Paladin, Grafton Books, London.

Paine, R. T. (1974) Intertidal community structure. Experimental studies on the relationship between a dominant competitor and its principal predator. *Oecologia (Berl.)*, **15,** 93–120.

Pamatmat, M. M. (1978) Oxygen uptake and heat production in a metabolic conformer (*Littorina irrorata*) and a metabolic regulator (*Uca pugnax*). *Mar. Biol.*, **48,** 317–25.

Parlour, J. W. and Schatzow, S. (1978) The mass media and public concern for environmental problems in Canada, 1960–1972. *Intern. J. Environ. Stud.*, **13,** 9–17.

Peakall, D. (1992) *Animal Biomarkers as Pollution Indicators*, Chapman & Hall, London.

Pearson, T. H. (1970) The benthic ecology of Loch Linnhe and Loch Eil, a sea-loch system on the west coast of Scotland. I. The physical environment and the distribution of the macrobenthic fauna. *J. Exp. Mar. Biol. Ecol.*, **5,** 1–34.

Pearson, T. H. (1971) The benthic ecology of Loch Linnhe and Loch Eil, a sea-loch system on the west coast of Scotland. III. The effect on the benthic fauna of the introduction of pulp mill effluent. *J. Exp. Mar. Biol. Ecol.*, **6,** 211–33.

Pearson, T. H. (1975) The benthic ecology of Loch Linnhe and Loch Eil, a sea-loch system on the west coast of Scotland. IV Changes in the benthic fauna attributable to organic enrichment. *J. Exp. Mar. Biol. Ecol.*, **20,** 1–41.

Pearson, T. H. (1980) Marine pollution effects of pulp and paper industry wastes. *Helgoländer Meeresunters.*, **33,** 340–65.

Pearson, T. H. and Rosenberg, R. (1978) Macrobenthic succession in relation to organic enrichment and pollution of the marine environment. *Oceanogr. Mar. Biol. A. Rev.*, **16,** 229–311.

Persoone, G. and Kihlström, J. E. (1988) Chairmen's report from session on ecotoxicological effects – testing and experimental methods. In: *First European Conference on Ecotoxicology Conference Proceedings*, (Eds H. Løkke, H. Tyle and F. Bro-Rasmussen), 17–19 Oct., 1988, Conference Organizing Committee, Lyngby, Copenhagen, Denmark. pp. 141–43.

Peters, R. H. (1991) *A Critique for Ecology*, Cambridge University Press, Cambridge.

Phillips, D. J. H. (1977) The use of biological indicator organisms to monitor trace metal pollution in marine and estuarine environments – a review. *Environ. Pollut.*, **13,** 281–317.

Phillips, D. J. H. (1980) *Quantitative Aquatic Biological Indicators*, Applied Science, London.

Poulsen, E., Riisgård, H. U. and Møhlenberg, F. (1982) Accumulation of cadmium and bioenergetics in the mussel *Mytilus edulis. Mar. Biol.*, **68,** 25–9.

Pratt, J. R. (1991) Making the transition from toxicology to ecotoxicology. In: *Organic Substances and Sediments in Water*, Vol. 3, (Ed R. A. Baker), pp. 25–42. Lewis, Chelsea, MI.

Pratt, J. R., Mitchell, J., Ayers, R. and Cairns, J., Jr. (1989) Comparison of estimates of effects of a complex effluent at differing levels of biological organization. In:

Aquatic Toxicology and Environmental Fate, 11th vol. ASTM STP 1007, (Eds G. W. Suter II and M. A. Lewis), pp. 174–88. American Society for Testing and Materials, Philadelphia.

Rand, G. M. and Petrocelli, S. R. (Eds) (1985) *Fundamentals of Aquatic Toxicology*, Hemisphere, New York.

Rapport, D. J. (1989) Symptoms of pathology in the Gulf of Bothnia (Baltic Sea): Ecosystem response to stress from human activity. *Biol. J. Linn. Soc.*, **37,** 33–49.

Rapport, D. J., Regier, H. A. and Hutchinson, T. C. (1985) Ecosystem behaviour under stress. *Am. Nat.*, **125,** 617–40.

Ray, D. R. (1991) Pesticides derived from plants and other organisms. In: *Handbook of Pesticide Toxicology*, Vol. 2, (Eds W. J. Hayes, Jr. and E. R. Laws, Jr.), pp. 585–636. Academic Press, San Diego, CA.

Reinfelder, J. R. and Fisher, N. S. (1991) The assimilation of elements ingested by marine copepods. *Science*, **251,** 794–6.

Rhoads, D. C. (1974) Organism–sediment relations on the muddy seafloor. *Oceanogr. Mar. Biol. Ann. Rev.*, **12,** 263–300.

Rhoads, D. C. and Boyer, L. F. (1982) The effects of marine benthos on physical properties of sediments: A successional perspective. In: *Animal–Sediment Relations: The Biogenic Alteration of Sediments*, (Eds P. L. McCall and M. J. S. Tevesz), 2nd vol., *Topics in Geobiology*, pp. 3–52. Plenum Press, New York.

Rhoads, D. C., McCall, P. L. and Yingst, J. Y. (1978) Disturbance and production on the estuarine seafloor. *Am. Sci.*, **66**, 577–86.

Rice, D. L. (1986) Early diagenesis in bioadvective sediments: Relationships between the diagenesis of beryllium-7, sediment reworking rates, and the abundance of conveyor-belt deposit-feeders. *J. Mar. Res.*, **44,** 149–84.

Rice, D. L. and Rhoads, D. C. (1989) Early diagenesis of organic matter and the nutritional value of sediment. In: *Ecology of Marine Deposit Feeders*, (Eds G. Lopez, G. Taghon and J. Levinton), pp. 59–97. Springer-Verlag, New York.

Rice, D. L. and Whitlow, S. I. (1985a) Early diagenesis of transition metals: A study of metal partitioning between macrofaunal populations and shallow sediments. In: *The Fate and Effects of Pollutants*, pp. 21–30. Maryland Seagrant Office, College Park, MD.

Rice, D. L. and Whitlow, S. I. (1985b) Diagenesis of transition metals in bioadvective sediments. In: *Heavy Metals in the Environment*, 2nd vol., pp. 353–5. CEC Consultants, Edinburgh.

Ricklefs, R. E. (1979) *Ecology*, 2nd edn, Chiron Press, New York.

Roesijadi, G., Drum, A. S., Thomas, J. T. and Fellingham, G. W. (1982) Enhanced mercury tolerance in marine mussels and relationship to low molecular weight, mercury-binding proteins. *Mar. Pollut. Bull.*, **13,** 250–3.

Roesijadi, G. and Fellingham, G. W. (1987) Influence of Cu, Cd, and Zn preexposure on Hg toxicity in the mussel *Mytilus edulis*. *Can. J. Fish. Aquat. Sci.*, **44,** 680–4.

Rosenberg, R. (1972) Benthic faunal recovery in a Swedish fjord following the closure of a sulphite pulp mill. *Oikos*, **23,** 92–108.

Rosenberg, R. (1973) Succession in benthic macrofauna in a Swedish fjord subsequent to the closure of a sulphite pulp mill. *Oikos*, **24,** 1–16.

Ross, J. B., Parker, R. and Strickland, M. (1991) A survey of shoreline litter in Halifax Harbour 1989. *Mar. Pollut. Bull.*, **22,** 245–8.

Roughgarden, J. (1977) Coevolution in ecological systems: Results from 'loop analysis' for purely density-dependent coevolution. In: *Measuring Selection in Natural*

Populations, 19th vol., *Lecture Notes in Biomathematics*, (Eds F. Christiansen and T. Fenchel), pp. 449–518. Springer-Verlag, New York.

Roughgarden, J., May, R. M. and Levin, S. A. (Eds) (1989) *Perspectives in Ecological Theory*, Princeton University Press, Princeton.

Sanders, B. (1990) Stress proteins: Potential as multitiered biomarkers. In: *Biomarkers of Environmental Contamination*, (Eds J. F. McCarthy and L. R. Shugart), pp. 165–91. Lewis Publishers, Boca Raton, FL.

Sanders, H. L., Grassle, J. F. and Hampson, G. R. (1972) The West Falmouth oil spill. I. Biology. *Woods Hole Oceanogr. Inst. Tech. Rep.*, 72–120.

Scala, R. A. (1991) Risk assessment. In: *Casarett and Doull's Toxicology: The Basic Science of Poisons*, 4th edn, (Eds M. O. Amdur, J. Doull and C. D. Klaassen), pp. 985–96. Pergamon Press, New York.

Schimmel, S. C., Garnas, R. L., Patrick, J. M., Jr. and Moore, J. C. (1983) Acute toxicity, bioconcentration, and persistence of AC 222,705, benthiocarb, chlorpyrifos, fenvalerate, methyl parathion, and permethrin in the estuarine environment. *J. Agric. Food Chem.*, **31**, 104–13.

Schindler, D. W. (1987) Detecting ecosystem responses to anthropogenic stress. *Can. J. Fish. Aquat. Sci.*, **44**, 6–25.

Schink, D. R. and Guinasso, N. L. (1978) Redistribution of dissolved and adsorbed materials in abyssal marine sediments undergoing biological stirring. *Am. J. Sci.*, **278**, 687–702.

Schink, D. R., Guinasso, N. L., Jr. and Fanning, K. A. (1975) Processes affecting the concentration of silica at the sediment-water interface of the Atlantic Ocean. *J. Geophys. Res.*, **80**, 3013–31.

Schlesinger, W. H. (1989) Discussion: Ecosystem structure and function. In: *Perspectives in Ecological Theory*, (Eds J. Roughgarden, R. M. May and S. A. Levin), pp. 268–74. Princeton University Press, Princeton.

Schluter, D. (1988) Estimating the form of natural selection on an quantitative trait. *Evolution*, **42**, 849–61.

Schumway, S. E., Scott, T. M. and Shick, J. M. (1983) The effects of anoxia and hydrogen sulphide on survival, activity and metabolic rate in the coot clam, *Mulinia lateralis* (Say). *J. Exp. Mar. Biol. Ecol.*, **71**, 135–46.

Seager, J. (1988) Deriving standards for water pollution control: The role of ecotoxicology. In: *Proceedings of the First European Conference on Ecotoxicology*, (Eds H. Løkke, H. Tyle and F. Bro-Rasmussen), 17–19 Oct., 1988, Conference Organizing Committee, Lyngby, Copenhagen, Denmark. pp. 383–403.

Shiaris, M. P. (1989) Seasonal biotransformation of napthalene, phenanthrene, and benzo(a)pyrene in surficial estuarine sediments. *Appl. Environ. Microbiol.*, **55**, 1391–9.

Shimp, R. J., Larson, R. J. and Boethling, R. S. (1990) Use of biodegradation data in chemical assessment. *Environ. Toxicol. Chem.*, **9**, 1369–77.

Sibly, R. M. and Calow, P. (1986) *Physiological Ecology of Animals: An Evolutionary Approach*, Blackwell Scientific, Oxford.

Sibly, R. M. and Calow, P. (1989) A life-cycle theory of responses to stress. *Biol. J. Linn. Soc.*, **37**, 101–16.

Silverman, B. W. (1986) *Density Estimation for Statistics and Data Analysis*, Chapman & Hall, London.

Simkiss, K. and Mason, A. Z. (1983) Metal ions: Metabolic and toxic effects. In: *The Mollusca*, vol. 2, pp. 101–64. Academic Press, New York.

Sipes, I. G. and Gandolfi, A. J. (1991) Biotransformation of toxicants. In: *Casarett and*

Doull's Toxicology: The Basic Science of Poisons, 4th edn, (Eds M. O. Amdur, J. Doull and C. D. Klaassen), pp. 88–126. Pergamon Press, New York.

Slobodkin, L. B. (1968) Aspects of the future of ecology. *Bioscience*, **18,** 16–23.

Sloof, W., Canton, J. H. and Hermens, J. L. M. (1983) Comparison of the susceptibility of 22 freshwater species to 15 chemical compounds. I. (sub)acute toxicity tests. *Aquat. Toxicol.*, **4,** 113–28.

Sloof, W., van Oers, J. A. M. and De Zwart, D. (1986) Margins of uncertainty in ecotoxicological hazard assessment. *Environ. Toxicol. Chem.*, **5,** 841–52.

Sokal, R. R. and Rohlf, F. J. (1981) *Biometry*, 2nd edn, W. H. Freeman, San Francisco.

Soulé, M. E. (1991) Conservation: Tactics for a constant crisis. *Science*, **253,** 744–9.

Sprague, J. B. (1985) Factors that modify toxicity. In: *Fundamentals of Aquatic Toxicology*, (Eds G. M. Rand and S. R. Petrocelli), pp. 124–63. Hemisphere, New York.

Squires, D. F. (1983) *The Ocean Dumping Quandary: Waste Disposal in the New York Bight*, State University of New York Press, Albany, NY.

Stegeman, J. J. (1980) Mixed-function oxygenase studies in monitoring for effects of organic pollution. *Rapp. P.-v Réun. Cons. Int. Explor. Mer*, **179,** 33–8.

Stephan, C. E. (1986) Proposed goal of applied aquatic toxicology. In: *Aquatic Toxicology and Environmental Fate*, 9th vol. ASTM STP 921, (Eds T. M. Poston and R. Purdy), pp. 3–10. American Society for Testing and Materials, Philadelphia.

Stephan, C. E., Mount, D. I., Hansen, D. J., Gentile, J. H., Chapman, G. A. and Brungs, W. A. (1985) Guidelines for deriving numerical national water quality criteria for the protection of aquatic organisms and their uses, PB85–227049, US Environmental Protection Agency, Washington DC.

Stephan, C. E. and Rogers, J. W. (1985) Advantages of using regression analysis to calculate results of chronic toxicity tests. In: *Aquatic Toxicology and Hazard Assessment: Eighth Symposium*, ASTM STP 891, (Eds R. C. Bahner and D. J. Hansen), pp. 328–8. American Society for Testing and Materials, Philadelphia.

Stern, A. M. and Walker, C. R. (1978) Hazard assessment of toxic substances: Environmental fate testing of organic chemicals and ecological effects testing. In: *Estimating the Hazard of Chemical Substances to Aquatic Life*, (Eds J. Cairns, Jr., K. L. Dickson and A. W. Maki), pp. 81–131. American Society for Testing and Materials, Philadelphia.

Sundelin, B. (1984) Single and combined effects of lead and cadmium on *Pontoporeia affinis* (Crustacea, Amphipoda) in laboratory soft-bottom microcosms. In: *Ecotoxicological Testing for the Marine Environment*, 2nd vol., (Eds G. Persoone, E. Jaspers, and C. Claus), pp. 237–58. State University of Ghent and Institute of Marine Scientific Research, Bredene, Belgium.

Suter, G. W. (1981) Ecosystem theory and NEPA assessment. *Bull. Ecol. Soc. Am.*, **62,** 186–92.

Swann, R. L., Laskowski, D. A., McCall, P. U. J., Vander Kuy, K. and Dishburger, H. J. (1983) A rapid method for the estimation of the environmental parameters octanol/water partition coefficient, soil sorption constant, water-to-air ratio, and water solubility. *Residue Rev.*, **85,** 17–28.

Swartz, R. C. (1987) Toxicological methods for determining the effects of contaminated sediment on marine organisms. In: *Fate and Effects of Sediment Bound Chemicals in Aquatic Systems*, (Eds K. L. Dickson, A. W. Maki and W. A. Brungs), pp. 183–98. Pergamon Press, Elmsford, NY.

Swartz, R. C., Schults, D. W., DeWitt, T. H., Ditsworth, G. R. and Lamberson, J. O. (1990) Toxicity of fluoranthene in sediment to marine amphipods: A test of the equilibrium partitioning approach to sediment quality criteria. *Environ. Toxicol. Chem.*, **9,** 1071–80.

Tagatz, M. E. and Ivey, J. M. (1981) Effects of fenvalerate on field- and laboratory-developed estuarine benthic communities. *Bull. Environ. Contam. Toxicol.*, **27,** 256–67.

Tagatz, M. E., Stanley, R. S., Plaia, G. R. and Deans, C. H. (1987) Responses of estuarine macrofauna colonizing sediments contaminated with fenvalerate. *Environ. Toxicol. Chem.*, **6,** 21–5.

Taghon, G. L., Self, R. F. L. and Jumars, P. A. (1978) Predicting particle selection by deposit-feeders: A model and its implications. *Limnol. Oceanogr.*, **23,** 752–9.

Tilman, D. (1989) Discussion: Population dynamics and species interactions. In: *Perspectives in Ecological Theory*, (Eds J. Roughgarden, R. M. May and S. A. Levin), pp. 89–100. Princeton University Press, Princeton.

Trevan, J. W. (1929) A statistical note on the testing of anti-dysentary sera. *J. Path. Bact.*, **32,** 127–34.

Trim, A. H. and Marcus, J. M. (1990) Integration of long-term fish kill data with ambient water quality monitoring data and application to water quality management. *Environ. Manag.*, **14,** 389–96.

Truhaut, R. (1977) Ecotoxicology: Objectives, principles and perspectives. *Ecotoxicol. Env. Safety*, **1,** 151–73.

Tucker, R. K. and Leitzke, J. S. (1979) Comparative toxicology of insecticides for vertebrate wildlife and fish. *Pharmacol. Therapeut.*, **6,** 220.

Underwood, A. J. (1989) The analysis of stress in natural populations. *Biol. J. Linn. Soc.*, **37,** 51–78.

Underwood, A. J. and Peterson, C. H. (1988) Towards an ecological framework for investigating pollution. *Mar. Ecol. Prog. Ser.*, **46,** 227–34.

Uthe, J. F., Freeman, H. C., Mounib, S. and Lockhart, W. L. (1980) Selection of biochemical techniques for detection of environmentally induced sublethal effects in organisms. *Rapp. P.-v. Réun. Cons. Int. Explor. Mer.*, **179,** 39–47.

Van Leeuwen, K. (1990) Ecotoxicological effects assessment in the Netherlands: Recent developments. *Environ. Manag.*, **14,** 779–92.

Van Leeuwen, C., Van der Zandt, P. T. J., Aldenberg, T., Verhaar, H. J. M. and Hermens, J. L. M. (1992) Application to QSARS, extrapolation and equilibrium partitioning in aquatic effects assessment. I. Narcotic industrial pollutants. *Environ. Toxicol. Chem.* 11, 267–82.

Van Straalen, N. M. and Denneman, G. A. J. (1989) Ecotoxicological evaluation of soil quality criteria. *Ecotoxicol. Environ. Safety*, **18,** 241–51.

Via, S. and Lande, R. (1985) Genotype–environment interaction and the evolution of phenotypic plasticity. *Evolution*, **39,** 505–22.

Via, S. and Lande, R. (1987) Evolution of genetic variability in a spatially heterogeneous environment: Effects of genotype–environment interaction. *Genet. Res., Camb.*, **49,** 147–56.

Volmer, J., Kördel, W. and Klein, W. (1988) A concept for environmental risk assessment of new chemicals. In: *Proceedings of the First European Conference on Ecotoxicology*, (eds H. Løkke, H. Tyle, and F. Bro-Rasmussen), 17–19 Oct., 1988, Conference Organizing Committee, Lyngby, Copenhagen, Denmark. pp. 411–6.

Wagner, C. and Løkke, H. (1991) Estimation of ecotoxicological protection levels from NOEC toxicity data. *Wat. Res.*, **25,** 1237–42.

Walker, C. H. (1980) Species variations in some hepatic microsomal enzymes that metabolize xenobiotics. *Prog. Drug Metabol.*, **5,** 113–64.

Wängberg, S.-Å., Heyman, U. and Blanck, H. (1991) Long-term and short-term arsenate toxicity to freshwater phytoplankton and periphyton in limnocorrals. *Can. J. Fish. Aquat. Sci.*, **48,** 173–82.

Warren, G. E. and Davis, G. E. (1967) Laboratory studies on the feeding bioenergetics and growth of fish. In: *The Biological Basis of Freshwater Fish Production*, (Ed S. D. Gerking), pp. 175–214. Blackwell Scientific, Oxford.

Warren, L. M. (1977) The ecology of *Capitella capitata* in British waters. *J. Mar. Biol. Assoc. UK*, **57,** 151–9.

Warwick, R. M. (1986) A new method for detecting pollution effects on marine macrobenthic communities. *Mar. Biol.*, **92,** 557–62.

Warwick, R. M., Pearson, T. H. and Ruswahyuni (1987) Detection of pollution effects on marine macrobenthos: Further evaluation of the species abundance/biomass method. *Mar. Biol.*, **95,** 193–200.

Watling, L. (1991) The sedimentary milieu and its consequences for resident organisms. *Am. Zool.*, **31,** 789–96.

Weis, J. S. and Weis, P. (1989) Tolerance and stress in a polluted environment: The case of the mummichog. *BioScience*, **39,** 89–95.

Weston, D. P. (1990) Hydrocarbon bioaccumulation from contaminated sediment by the deposit-feeding polychaete *Abarenicola pacifica*. *Mar. Biol.*, **107,** 159–69.

Whitlatch, R. B. (1974) Food-resource partitioning in the deposit-feeding polychaete *Pectinaria gouldii*. *Biol. Bull.*, **147,** 227–35.

Whittaker, R. H. (1975) *Communities and Ecosystems*, 2nd edn, Macmillan, New York.

Widdows, J. and Hawkins, A. J. S. (1989) Partitioning of rate of heat dissipation by *Mytilus edulis* into maintenance, feeding, and growth components. *Physiol. Zool.*, **62,** 764–84.

Widdows, J. and Johnson, D. (1988) Physiological energetics of *Mytilus edulis*: Scope for growth. *Mar. Ecol. Prog. Ser.*, **46,** 113–21.

Williams, G. C. (1975) *Sex and Evolution*, Princeton University Press, Princeton.

Williamson, M. H. (1987) Are communities ever stable? In: *Colonization, Succession and Stability*, (Eds A. J. Gray, M. J. Crawley and P. J. Edwards), pp. 353–71. Blackwell Scientific, Oxford.

Wilson, M. V. and Botkin, D. B. (1990) Models of simple microcosms: Emergent properties and the effect of complexity on stability. *Am. Nat.*, **135,** 414–34.

Windsor, J. G., Jr. and Hites, R. A. (1979) Polycyclic aromatic hydrocarbons in Gulf of Maine Sediments and Nova Scotia soils. *Cosmochim. Geochim. Acta*, **43,** 27–33.

Winberg, G. C. (1956) Rate of metabolism and food requirements of fishes. *Fish. Res. Bd. Can. Transl. Ser. No.* **194,** 1–253.

Wolff, W. J. and Zijlstra, J. J. (1980) Management of the Wadden Sea. *Helgoländer Meeresunters.*, **33,** 596–613.

Wood, R. J. (1981) Insecticide resistance: Genes and mechanisms. In: *Genetic Consequences of Man Made Change*, (Eds J. A. Bishop and L. M. Cook), pp. 53–96. Academic Press, London.

Wood, R. J. and Bishop, J. A. (1981) Insecticide resistance: Populations and evolution. In: *Genetic Consequences of Man Made Change*, (Eds J. A. Bishop and L. M. Cook), pp. 97–128. Academic Press, London.

Woodwell, G. M. (1991) Supporting life on earth. (Letter), *Science*, **254,** 358–9.

Woonacott, T. H. and Woonacott, R. J. (1990) *Introductory Statistics*, 5th edn, John Wiley and Sons, New York.

Index